The Biopolitics of Water

Biopolitics refers to a form of politics concerned with administering and regulating the conditions of life at an aggregated level of populations. This book provides a biopolitical perspective on water governance and its effects. It draws on the work of Foucault to explore how notions of scarcity are used in strategies of governance and how such governance differentiates between different populations. Furthermore, the author investigates what such biopolitical regulation means for people's lifestyles and the way they understand themselves and their moral responsibilities as humans, individuals, and citizens.

The book begins by investigating the global water agenda, with a particular emphasis on its focus on water for basic needs, and provides different examples of hydromentalities around the world. It also presents rich empirical details of one local case in South Africa. By carefully exploring the water 'stories' of water users, the book provides new perspectives on the relationship between water and power. Additionally, it offers an innovative methodological framework through which we can study the workings of governance more generally, and water governance specifically. It thereby contributes to the scholarship on water governance in relation to how water governance and technologies are part of producing subjectivities, notions of life and lifestyles and, more specifically, how the global water agenda can work so as to produce, or further entrench, distinctions between different lives and lifestyles. Ultimately, differences between individuals and populations that are produced as an effect of water governance are assessed in relation to *social* sustainability.

Sofie Hellberg is a Postdoctoral Researcher at the University of Gothenburg, Sweden and the University of the Witwatersrand, South Africa.

Earthscan Studies in Water Resource Management

Transboundary Water Governance and International Actors in South Asia
The Ganges–Brahmaputra–Meghna Basin
Paula Hanasz

The Grand Ethiopian Renaissance Dam and the Nile Basin
Implications for transboundary water cooperation
Edited by Zeray Yihdego, Alistair Rieu-Clarke and Ana Elisa Cascão

Freshwater Ecosystems in Protected Areas
Conservation and management
Edited by Max C. Finlayson, Jamie Pittock and Angela Arthington

Participation for Effective Environmental Governance
Evidence from European Water Framework Directive Implementation
Edited by Elisa Kochskämper, Edward Challies, Nicolas W. Jager and Jens Newig

China's International Transboundary Rivers
Politics, Security and Diplomacy of Shared Water Resources
Lei Xie and Jia Shaofeng

Urban Water Sustainability
Constructing Infrastructure for Cities and Nature
Sarah Bell

The Biopolitics of Water
Governance, Scarcity and Populations
Sofie Hellberg

For more information and to view forthcoming titles in this series, please visit the Routledge website: www.routledge.com/books/series/ECWRM/

The Biopolitics of Water

Governance, Scarcity and Populations

Sofie Hellberg

Routledge
Taylor & Francis Group
LONDON AND NEW YORK

earthscan
from Routledge

First published 2018 by Routledge

2 Park Square, Milton Park, Abingdon, Oxon OX14 4RN
605 Third Avenue, New York, NY 10017

Routledge is an imprint of the Taylor & Francis Group, an informa business

First issued in paperback 2021

Publisher's Note

The publisher has gone to great lengths to ensure the quality of this reprint but points out that some imperfections in the original copies may be apparent.

British Library Cataloguing-in-Publication Data
A catalogue record for this book is available from the British Library

Library of Congress Cataloging-in-Publication Data
A catalog record has been requested for this book

ISBN: 978-1-138-74075-4 (hbk)
ISBN: 978-0-367-50839-5 (pbk)

Typeset in Sabon
by Wearset Ltd, Boldon, Tyne and Wear

To my sons Hannes and Hjalmar and their father, my partner, Beniamin.

Contents

Acronyms and abbreviations

CBD	Convention on Biological Diversity (United Nations)
EBM	Ecosystem-based Management
EWS	eThekwini Water and Sanitation
FBW	Free Basic Water
GWG	Global Water Governance
GWP	Global Water Partnership
IDP	Integrated Development Plan
IWA	International Water Association
IWRM	Integrated Water Resources Management
LCPD	Litres per Capita Per Day
MDG	Millennium Development Goal
MT	Mvula Trust
NGO	Non-Governmental Organization
NP	Nationalist Party (South Africa)
POAT	Plan de Ordenamiento Ambiental Territorial/Environmental zoning plan for watersheds (Paraguay)
PPP	Public–Private Partnerships
RDP	Reconstruction and Development Programme
SDG	Sustainable Development Goal
SNWT	South–North Water Transfer (China)
UN	United Nations
UNDP	United Nations Development Programme
UNEP	United Nations Environmental Programme
UNESCO	United Nations Educational, Scientific and Cultural Organization
WHO	World Health Organization
WRG	Water Resources Group
WRI	World Resources Institute
WWDR	World Water Development Report
ZAR	South African Rand

1 The power and politics of water

> The emotion of water and the access to it ... I have had women who put their crying baby on the phone. I have had women hysterical. I have had people coming in here with guns and threatened to shoot me. I have had staff shot at and killed for water. So I know when you deprive somebody of water how emotionally.... There is this thing in you: it is like you close off somebody's air supply. That fight for water and air. Because five minutes with no air and five days without water you are dead. Five weeks with no food. So the famous five.... But it is the minutes and the days, you know, in five days: if I haven't had water I am dead. That triggers something in people, some irrationality. So it's highly emotive. And people play on that. So you've got a very strong emotion that you could use. It's like a feeling of injustice almost. So the manipulative, the anarchists, the clever ones try and play on that. I am in this game and I have seen it.

The above quotation is an excerpt from an interview with the then (2009) Head of Water and Sanitation (EWS) in eThekwini municipality, South Africa, and it is a good illustration of some of the main themes of this book. The quotation revolves around strong emotions that surround water because it is a necessity for *survival*. It highlights the *contestations* and tensions around its distribution. It illustrates, through the example of hysterical women, how water is *gendered*. It also tells a particular story of how emotions – despite the fact that they are stirred by the fight for survival – are associated with irrationality. It indicates how *resistance* against perceived injustices in water distribution are seen as something that is used strategically for *political* purposes. Finally, it signals that what the EWS stands for, in contrast to the emotive and political, is the *rational*.

In this book the representation of water and its functions, and its management and allocation, are addressed as inherently political and, more specifically, biopolitical. This means that water management systems are critically scrutinized in terms of their attempts at governing life and the distinctions they make between different forms of lives, as well as what such ways of governing mean for people's lifestyles and how they understand themselves and their (moral) responsibilities as humans/individuals/citizens.

The statement above is indicative of a particular understanding of water and its proper management. Such specific definitions of water and water governance in this book are understood as forming a 'hydromentality'. Hydromentalities are assemblages of governing rationalities, techniques of ruling, and ways of thinking about and defining water, which govern both water use and water users. One of the aspects of a hydromentality is, as Linton (2010) has argued in relation to the hegemony of 'modern water', that it designates 'a particular kind of identity, representation, and material form of water'. As such, it does not acknowledge other ways of 'imagining, representing and materializing water' as they are 'made out to be less real or less legitimate' (ibid., p. 11). Hence, they are viewed as irrational.

Recognizing biopolitical aspects of water governance means opening up the box of what water is and what water does. The answers to these questions are not as self-evident as they might first seem (Linton, 2010). The biopolitical perspective this book applies will help us look into the way that water and water regulation are part of a politics that governs individuals and populations in terms of their conditions of life (see Dean, 1999; Foucault, 1998). In the course of which water is addressed as an element that is productive for our ways of being humans and for the societies in which we live. Hence, water is not only a resource necessary for survival but 'what we make of it' produces 'the worlds and the selves we inhabit' (Linton, 2010, p. 3). Water can thus be understood as exceptional in that it is the prerequisite for all manner of life, as well as for *how* these lives are lived. Water can, however, also be seen as *un*exceptional, as its everydayness makes it 'so normal that it retreats into the background of awareness' (Sofoulis, 2005, p. 448). It is around the intersection between this exceptionality and the everydayness of water that this book revolves.

The politics of water during the last decades has been increasingly addressed under the rubric of 'hydropolitics' (Turton, 2002; Waterbury, 1979). Decisions in hydropolitics are concerned with the questions of 'who gets what [water], when, where and how' (Turton, 2002, p. 16). When studying these processes, previous research – especially in critical geography and under the wider umbrella of political ecology – has argued for the recognition of how power relations in water management produce, redistribute, and determine access to, or exclusion from, water (Loftus, 2009; Swyngedouw, 1999, 2004). Thus, for Swyngedouw, 'the water problem is not merely a question of management and technology, but rather, and perhaps in the first instance, a question of social power' (Swyngedouw, 2004, p. 175). Such critical research has analysed neoliberal trends in water management, including the effects of privatization, cost recovery policies, and water pricing systems (Bakker, 2005, 2007, 2010; Barlow and Clarke, 2003; McDonald and Pape, 2002; McDonald and Ruiters, 2005; Swyngedouw, 2005). By focusing on power relations, and both the discursive and the material underpinnings of water regulation, critical researchers have politicized 'hydrosocial' relations in a number of ways. They have focused on the ways in which water acquires certain meanings and how these meanings and representations of water are pivotal for understanding 'who gets access to it and on what terms' (Linton, 2010, p. 69; see also Johnston, 2003; Linton and Budds, 2014; Loftus, 2015; Shiva, 2002; Strang, 2004, 2009).

As mentioned, this book explores hydropolitics as a form of biopolitics. The way that biopolitics is defined and understood here (outlined in more detail in Chapter 2) invites an exploration of how such governance constructs populations and problematizes and governs their use of water resources. In turn, part of the methodological set-up of this book is to explore how such governing strategies are productive of lifestyles and identities, and how they relate to ideas of what it is to live a good and fulfilling life, as narrated by the water users themselves. These are no small matters. To quote Judith Butler, such a question of 'what makes for a livable world is no idle question' (2004, p. 17). It is not only a question of how one views one's own life, but also a question of 'what makes, or ought to make, the lives of others bearable'. In turn, finding answers to these questions, Butler argues: 'we find ourselves not only committed to a certain view of what life is, and what it should be, but also of what constitutes the human, the distinctively human life, and what does not' (ibid.).

What is at stake regarding the relationship between water and life is, hence, not merely survival. It is about what is recognized as a decent and proper life. The complexities of these questions as regards what is appropriate for different populations come to the fore in water governance, understood in this book as a form of biopolitical governance that is based on a population discourse that distinguishes between different forms of lives.

Such complexities are also prominent in the water users' stories. In these narratives, the (bio)political claim that everyone should have the right of access to basic water is straightforwardly acknowledged while at the same time being imbued with ambivalence. These uncertainties relate to how water users understand their own right in relation to other people's right to water, making this into a question of the distinctions made between Self and Other. These ambivalences also revolve around the notion of payment as the (just) way of distributing water above the threshold of survival.

From a more theoretical perspective, the right to water for survival raises questions regarding whether a claim to such a right is to be understood as a simultaneous claim to political recognition, or whether the demand for the right to survival should be conceptualized as part of an assemblage that attempts to keep those who are 'simply alive' (cf. Agamben, 1998) from the 'good life'. These ambivalences in relation to the notion of life are symptomatic of (neo)liberalism and its ways of simultaneously making (bare) life political and making a distinction between different forms of lives, hence both including and excluding life (understood as *zoë*, biological life) in (relation to) political existence (understood as *bios*, proper or qualified life) (cf. Agamben, 1998).

Although this book does not attempt to give a definite answer to these questions, which relate to the debates on the different understandings of biopolitics as conceptualized by Foucault and Agamben, let alone settle the debates, they are at the heart of book. A biopolitical analysis, such as the one conducted in this book, also acknowledges the distinctions between animals and humans, even though they are a less prominent theme. The book, furthermore, offers a perspective on the complicated task of conceptualizing agency (including resistance) in relation to the strategies of a government that uses the very freedom of its subjects to govern. This first chapter provides an overview of critical research that focuses on the power and politics of water and its governance. It also introduces how a biopolitical perspective on water governance adds to our understanding of the complexities of the politics of water, while outlining the whole book.

The politics of water: from hydraulic societies to hydromentalities

The idea of relating water (and its regulation) to certain kinds of societies not new. Wittfogel, in his classic work *Oriental Despotism* (1957), outlined how large-scale irrigation systems, and the centralized bureaucracy needed to maintain them, produced despotic and powerful regimes, which he termed 'Hydraulic societies' (Linton, 2010, p. 64; Strang, 2004, p. 21; Wittfogel, 1957). Although Wittfogel's hypothesis has been widely contested (see, for example, Barker and Molle, 2004; Davies, 2009; Molle, Mollinga and Wester, 2009), his 'approach to the relationship between society and water (nature) might ... be considered something of a forerunner to more recent historical and cultural investigations into what we are calling the "social nature of water"', as Linton has put it (2010, p. 64). Through what has been termed the 'hydrosocial cycle' (see Heynen *et al.*, 2005; Linton, 2010; Linton and Budds, 2014; Loftus, 2015; Swyngedouw, 2009, 2015), such research has studied the coproduction (cf. Linton, 2010) between water and society. Swyngedouw explains:

> In a sustained attempt to transcend the modernist nature–society binaries, hydro-social research envisions the circulation of water as a combined physical and social process, as a hybridized socio-natural flow that fuses together nature and society in inseparable manners.
>
> (Swyngedouw, 2006, cited in Swyngedouw, 2009, p. 56)

According to such approaches, 'society and water can be understood to make each other' (Linton, 2010, p. 5). This is a fundamental insight developed in this book. Below I specifically focus on three main ways of approaching the politics and power of water that are specifically relevant to the development of the argument of this book: how the global water crisis can be perceived as constructed; how unequal power relations in water allocation have been addressed; and how researchers have interrogated the relationship between water, water technologies, and subjectivities/citizenship. This last point specifically addresses the way in which water has been understood as biopolitics, which is a theme developed throughout this book.

(Global) water problems as constructed

Taking a critical stance on how water and water use are governed requires looking into how water problems are constructed in the first place. This does not mean to say there are no water problems in the world, but that the ways in which they are represented fit into larger narratives of resource and environmental governance. In these narratives, problems and solutions are framed in certain ways due to particular political and social circumstances (Linton, 2010, p. 7).

Current policy discourses on water are centred on the Integrated Water Resources Management framework (IWRM).[1] IWRM marks a shift from previous hydraulic and scientific discourses, which were characterized by a belief in an abundant water supply that could meet the growing needs of a modernizing society through hydraulic technologies (Bakker, 2010). This shift is more closely explored in Chapter 4. Here, suffice it to note that IWRM is described as an integrated approach that takes into account the dimensions of *sustainability* and *equity* along with *efficiency*. IWRM is thus closely connected to the mainstream definition of sustainable development in terms of its three pillars (environmental, social, economic). According to the IWRM approach, a more conserving and integrated way of managing water is needed, both because of decreasing availability and to provide access for those who are lacking water services. The approach has been immensely influential and is subsequently seen as the way forward. IWRM 'has become *the* discursive framework of international water policy – the reference point to which all other arguments end up appealing' (Conca, 2006, cited in Linton, 2010, p. 216, italics in original). As global water governance has evolved, however, new frameworks and concepts have emerged. Among those, water security and the nexus approaches (food–water–energy) are the most prominent. Their relation to the IWRM approach is discussed in Chapter 3, which concludes that IWRM is still the overarching framework for global water governance.

In Linton's (2010) analysis of the construction of the so-called global water crisis that subsequently paved the way for the IWRM framework, Gleick's edited volume *Water in Crisis* (1993) is viewed as one important factor that led to water becoming understood in a new way. In Gleick's publication, a gloomy picture of the world's water situation was painted:

> As we approach the 21st century we must now acknowledge that many of our efforts to harness water have been inadequate or misdirected. We remain *ignorant of the functioning of basic hydrological processes*. Rivers, lakes and ground water aquifers are increasingly *contaminated* with biological and chemical wastes. Vast numbers of people *lack clean drinking water* and rudimentary sanitation services. *Millions of people die every year* from water related diseases such as malaria, typhoid and cholera. *Massive water developments have destroyed* many of the world's most productive wetlands and other aquatic habitats.
>
> (1993, p. 3, my italics)

Thus, Gleick brought attention both to previous destructive water allocation strategies and to problems to do with lack of availability, and pollution and its devastating consequences. These issues were understood to be mainly *scientific problems*, as Gleick hints at when mentioning that we are ignorant of the functioning of the hydrological processes. This becomes even clearer in Shiklomanov's second chapter:

> The *rational use* and protection of water resources and supplying humankind with adequate clean fresh water, or, to be concise, 'water problems,' are among today's most acute and complex *scientific and technical problems*. They increasingly reach beyond national and regional borders and are becoming global in nature. Shortages of freshwater and the increasing pollution of water bodies are becoming limiting factors in the economic and social development of many countries throughout the world, even countries not located in arid zones.
>
> (1993, p. 13, my italics)

He continues:

> Under these conditions the reliable assessment of water resources is extremely important, *particularly the quantitative estimation and calculation of annual stream flows* and their fluctuations in time and space. This is due to the fact that the mean annual river runoff and annually renewable ground water resources almost everywhere support the bulk of water consumption and *determine the availability of water supply* for a given region.
>
> (Ibid.)

The point here is that the global water crisis, as it became known in the 1990s, was framed within the paradigm of what Linton (2010) calls 'modern water'. In this paradigm, questions of water are rendered scientific and technical, and the amount of water that can be made available for human use is presented as a fixed quantity – a natural fact. According to

Linton, the crisis was constructed through a description of a static relationship between the availability of water and population(s) (Linton, 2010, p. 193). As such, the crisis did not in any deep sense challenge the paradigm of modern water. Challenging this paradigm, Linton argues, would involve to a greater extent a changing relationship to water and how it is used. Moreover, Linton is critical of how the water crisis is portrayed as taking place on a global scale:

> to gather problems of water under the rubric of global water crisis diverts attention from the political and social circumstances that produce such problems and frames their solution in predominantly technical and hydrological terms.
>
> (2010, p. 7)

Critical scholars, mainly within political ecology, have studied the social production of the water crisis and the 'manufacturing of scarcities' (Johnston, 2003; Linton, 2010, p. 68). Within such frameworks, the water crisis is not something that is natural or neutral, but rather something that is socially constructed within the present social order. Swyngedouw offers a different image of water problems compared to that of Gleick and Shiklomanov:

> True scarcity does not reside in the physical absence of water in most cases, but in the lack of monetary resources and political and economic clout. Poverty and governance that marginalizes makes [sic] people die of thirst, not absence of water.
>
> (2009, p. 58)

Loftus argues similarly, in relation to the concept of water security – a concept which has recently been incorporated in global water governance – that often the causes of water insecurity are naturalized. He writes: 'this naturalising tendency quickly slips into a form of determinism: water insecurity is the result of insufficient rain rather than the unjust distribution of supplies of potable water' (Loftus, 2015, p. 351).

Swyngedouw further contends that 'nature and its waters do not exist outside the metonymic chains and social practices that offer some sort of instable meaning of what nature or water is' (2015, p. 228) and that:

> [a]ll socio-spatial processes are invariably predicated upon the circulation, metabolism, and summoning of particular social, cultural, physical, chemical, or biological processes, and their outcome is contingent, often unpredictable, and always immensely varied and risky.... Processes of metabolic change are, therefore, never socially or ecologically neutral.
>
> (Ibid.)

Taking these lines of critique as a point of departure, throughout this book I examine the meanings and definitions given to water, in particular historical and geographical contexts, as well as how ideas of physical access to water and notions of *scarcity* work so as to support a particular form of governance. I now turn to a related theme within the politics of water.

Unequal power relationships and distribution of water

Critical scholars of water governance have focused on inequalities in relation to water management, access, and use. By scrutinizing the effects of commodification, liberalization, and privatization trends in water management, researchers have made water struggles in disadvantaged communities a central research area (see, for example, Barlow and Clarke, 2003; Birkenholtz, 2009; Castro, 2008; Loftus, 2005b; McDonald and Pape, 2002; McDonald and Ruiters, 2005; Shiva, 2002; Sultana and Loftus, 2012; Swyngedouw, 2005).

Not only has this research theme been prominent in relation to cases such as the Cochabamba 'water war'[2] (Oliviera, 2004) – which has served to demonstrate the negative aspects of privatization, as well as the dynamics and possibilities of resisting such neoliberal governance – but it has also challenged water governance conventionally understood as more inclusive and pro poor. For example, Birkenholtz has recently shown that in Jaipur, India, socio-economic difference is being exacerbated even with a progressive water policy (2010, p. 2250), which has also been displayed in the case of eThekwini (see, for example, Hellberg, 2014, 2017; Loftus 2005a, 2005b, 2006, 2007, 2009), a case to which we will return.

In relation to unequal power relations in water governance and management, 'gender and water literature' (Ahlers and Zwarteveen, 2009; Harris, 2008, 2009; O'Reilly, 2006; Sultana, 2009) is important. This literature not only highlights the differentiated effects of changes in water access in terms of their gendered aspects, but it also offers a broader understanding of the production of subjectivity and of how intersectionality functions in relation to water and water access. Such gender analyses, as Harris (2008, p. 2644) points out, are 'essential to understand uneven consequences of water quality or access'. Given the focus on intersectionality, such analyses call for particular attention 'to *diverse and interlinked* processes of socio-economic differentiation ... as co-constitutive, rather than as analytically separable' (ibid., italics in original).

In the context of southern Africa, research has highlighted how cost recovery policies severely affect affordability and access to services in poor communities (McDonald and Pape, 2002) and how the more general trends of commodification and privatization of water and water management have affected access to services in a number of cases, ranging from Cape Town to Lusaka (McDonald and Ruiters, 2005). McDonald and Ruiters warn that these trends: 'might cement old-style apartheid and class geographies and further reinforce discriminatory aspects of infrastructure distribution that characterized colonialism and apartheid in the region, making a meaningful transition to more egalitarian social structure unlikely' (2005, p. 9).

South African water politics are a recurring theme in this book. It is important to acknowledge that my engagement with the biopolitics of sustainable development in general, and that of water governance in particular, stems from an interest in understanding the dynamics in South Africa. In this context, questions of water availability and accessibility are (and have long been) both extremely unequal and highly politicized.

However, at the same time as noting that seeing water governance in this particular place has affected how water governance is conceptualized, it is important to recognize that South Africa is not just any case, as understood in relation to global policy prescriptions. According to global standards South Africa's water policies have been showcased as a success story in the advancement of integrated water management practices (see for example WWDR, 2006, p. 13). Above all, the fact that the South African constitution defines 'sufficient water' as a right (RSA, 1996), implemented through the policy of Free Basic Water (FBW), has gained global attention as a way forward for promoting fairer access to water. The challenges that the country is facing can, furthermore, be seen as mirroring the projected image of the conditions regarding water on a global scale; in South Africa, as well as globally, water scarcity is looming in a situation where many urgently need extended access to services in a context of high inequality and uncertainty due to climate change. How (unequal) power relations play out in the South African context is examined in Chapters 4 and 5 and is a central theme throughout this book. One of the dimensions of power relations are their productive effects and the following shows how this has been addressed in previous research.

Productive aspects of water governance/water technologies

Studies of the politics of water have drawn attention to how the materiality of water and water technologies mediate water–society relations in that they are productive of certain meanings of, and relationships to, nature, modernity, and the city, as well as giving meaning to specific places (see, for example, Gibbs, 2006, 2009; Kaika, 2005; O'Reilly, 2006; Swyngedouw, 2004, 2015).

In this tradition, studies have focused on the dialectic relationship between water and society and how the materiality of water and water technologies – which themselves are 'social technical and socio natural' (Bakker, 2012, p. 616) – play a role in 'shaping human perceptions, discursive constructions, and responses to water' (ibid., p. 617). In line with such research, this book addresses technology as a central component of how the political dimension of water allocation is understood and studied, the different devices being part of the 'political calculability' (Braun and Whatmore, 2010, p. xi) for delivering water services.

For instance, according to Kaika, technology networks can be understood as 'the carriers of goods and services that promise to deliver the modernist dream for a happier, better society, with supply of water, health and food for all' (2005, p. 33). Furthermore, technology networks can be seen as 'carriers of modernist promise of participating in the phantasmagorical new world of technological advancement and progress; a world in which human freedom and emancipation resides in connecting to technology' (ibid.). With specific regard to dams, they can be portrayed as grand expressions of 'human power and will' (Sofoulis, 2005, p. 452). However, they can also be depicted as 'spectacular concretisations of colonial and/or dependency relations' (ibid.).

Moreover, research has elaborated on how we can understand the interaction between water users and water infrastructure (both big water and domestic infrastructure) as productive of notions of hygiene, cleanliness, and self-presentation (see Bakker, 2012; Gandy, 2004; Shove, 2003; Sofoulis, 2005). Thus, water infrastructure systems carry ideas of modernity, and expectations of access to services and of a better (and proper way of) life. This is explored through the local example of eThekwini in Chapter 5 of this book.

Recent work in relation to this strand of research has placed an emphasis on how interactions with water and water technologies can be constitutive of identities and how they are bound up with specific notions of citizenship (Birkenholtz, 2008, 2009, 2013; Ekers and Loftus, 2008; Kooy and Bakker, 2008a, 2008b; O'Reilly, 2006; Rodina and Harris, 2016; Strang, 2004; Sultana, 2009, 2011, 2013; von Schnitzler, 2008).

In South Africa specifically, research on water technologies has highlighted how the technology applied to the allocation of water services can be understood as symbolic of the relationship between democratic South Africa and its citizens. Von Schnitzler (2008) problematizes the way that prepayment water meters signal the possibility of being connected to the water network and getting access to free basic water, but, simultaneously, requiring that the water user become individually responsible for optimizing their household's consumption. Furthermore, von Schnitzler problematizes the fact that the implementation of these technological devices asks users to become 'economising actors' through 'calculative agencies'. Thus, according to von Schnitzler, we can understand inclusion in water service delivery, and in the state more generally, as being 'contingent upon a successful performance of an ethic that fuses civic duty and entrepreneurial comportment' (ibid., p. 916).

This book aims to build on this research and to continue interrogating the relationship between water, water technologies, and citizenship/subjectivities from a biopolitical perspective. Next, we turn to Foucauldian approaches to the study of water.

Foucauldian approaches

While Neo-Marxian and Gramscian approaches to analysing the materiality of water and water politics remain dominant in the literature, some work on water governance draws explicitly on Foucauldian power theory and notions of (green) governmentality and biopolitics to address power relations and modes of governing in water politics and management (Babu, 2009; Birkenholtz, 2009; Boelens *et al.*, 2015; Ekers and Loftus, 2008; Gandy, 2006; Hansson, 2014; Hellberg, 2014, 2017; Joyce, 2003; Kooy and Bakker, 2008b; Rogers *et al.*, 2016; Ward, 2013).

Such research has addressed the relationships between technological networks and human bodies in urban space, and how these relationships can be problematized and disentangled (Gandy, 2004, 2006, 2014); and how water supply infrastructure in urban space is constitutive of different identities and notions of particular modern citizens (Joyce, 2003; Kooy and Bakker, 2008b). In their description of colonial and contemporary Jakarta, Kooy and Bakker draw on notions of both governmentality and materiality and argue that water infrastructure, both historically and currently, has played/is playing a role in physically facilitating a division of populations and urban spaces (2008b, p. 385). They outline how discursive regimes around water supply systems in colonial times have shifted from being based on racial segregation to revolving instead, in the post-colonial era, around class-based markers of status (ibid., pp. 376, 385), and how access to 'modern' piped water is connected to one's identity as a modern citizen. Here, 'modern' water is signified (in part) by clean water, privatized spaces for water use, and understandings of water's economic cost/benefits (ibid., p. 385).

Other studies that draw explicitly on the notion of governmentality (and that are discussed further in Chapter 4), show how specific forms of governance are employed in policy and practice in different national and local contexts. Among other things, these studies provide examples of how participation, education, and awareness campaigns are used to transform the perceptions of water users (Babu, 2009; Birkenholtz, 2009; see also Chapter 5 in this book). These studies also give insights into how subjectivities, both of water professionals (Hansson, 2014; Ward, 2013) and of water users, are constructed (Hellberg, 2014, 2017), as well as focusing on strategies of resistance (Birkenholtz, 2009; Boelens *et al.*, 2015; see also Chapter 5 in this book).

Bakker has developed Foucauldian notions further in water-related research by addressing water as biopolitics. According to Bakker, water can be understood as biopolitical since 'modern governments seek to optimize both water resources and our individual water-use practices in order to secure the health and productivity of the population' (in Bakker, 2010, p. 190 and 2012, p. 619). In Bakker's work, this biopolitical element of water 'connects individual bodies to the collective body politic', as water is the carrier of bodily wastes, vectors of disease and pollution, which require regulation and control (ibid.).

In conversation with scholars, such as Gandy (2006) and Legg (2006), who focus on the infrastructure, biopolitics, and technologies of government in both the Global North and the Global South, and with those scholars, such as Bashford (2006) and Goubet (1989), who have elaborated on the role of modern hygiene practices for identity construction, Bakker contends that water governance is a form of biopolitics that is based on 'the categorization, quantification, and knowledge/power formation of urban residents in an attempt to govern their behaviour' (Bakker, 2013, p. 283).

Furthermore, building on critical and postcolonial development scholars, such as Appadurai (2001), Chatterjee (2004), Kabeer (2002), and Mamdani (1996), Bakker argues that it is important to analyse water as biopolitics because it is useful for unsettling conventional perspectives on water and development in postcolonial contexts (Bakker, 2013). Such perspectives view the extension of public services as progressive. However, in a post-colonial context, where exclusion from public services, and thus unevenness, have been integral to processes of modernization, such perspectives might not apply. Questions of resource availability, issues of identity, and spatial fragmentation, where previously only a fragment of the population has been regarded as citizens, all play a role in complicating a universal inclusion of service delivery. Furthermore, building on Tania Murray Li (2007), Bakker notes that conventional approaches to development involve processes that 'render technical' (Bakker, 2013) and depoliticize issues of service delivery. A biopolitical perspective, she contends, can help make explicit the dialectic between inclusion/exclusion that service delivery entails, and furthermore, make sense of the relationship between the governing and the governed in relation to the way that governmentalities unfold.

In this book these insights into the biopolitical nature of water, as well as its usefulness in unsettling conventional understandings of water service delivery and processes of development, are taken as a point of departure. I elaborate on the notion of biopolitics in terms of how it is relevant in an environmental and water governance context, and how we can understand water governance as biopolitical as it takes shape in global governance, and on the biopolitics of water in a particular local context. I now introduce how the notions of governmentality and biopolitics are applied and what we can learn from this approach in terms of the politics and power of water.

The biopolitics of water and hydromentalities

The Foucauldian understanding of government, governmentality, and biopolitics is a central part of the analytical framework of this book, as is his view on neoliberalism. These concepts are presented in more detail in Chapter 2, but here I will briefly introduce some of the most important aspects of these concepts.

As regards the first concept, *government*, it is important to note that it should be understood as distinct from *governance*. Governance, in this context, is understood as referring to a strategy, or process, for regulating and managing problems in which it is assumed that a neutral stance towards political decisions can be attained (Lemke, 2007, p. 55). Furthermore, such strategies and processes of governance involve a preoccupation with 'the search for ' "rational", "responsible" and "efficient" instruments of problem management' (ibid.). An engagement with the concept of government, in the Foucauldian sense, in contrast, focuses on the ways in which governing not only responds to 'reality' but is also active in producing it, through the constitution of agents and their interests, of identities and subjectivities, and, furthermore, on the close connection between technologies of rule and politics and how regimes of power/knowledge legitimate particular forms of governance (ibid.).

The second concept, *governmentality*, is employed in this book to analyse how government is put to work through 'organized practices', involving a variety of 'mentalities, rationalities, and techniques' (Mayhew, 2004) in different historical and geographical contexts.

The third concept, *biopolitics*, refers to the regulation of a population's life conditions (Dean, 1999; Foucault, 1998, p. 139, 2003). Such biopolitical governance is a central aspect of a Foucauldian conceptualization of government in contemporary modern societies. In such descriptions biopolitical strategies for governing are a crucial and necessary part of *neoliberal* governmentality (Dean, 1999, p. 113).

Neoliberalism, the fourth concept, should be understood somewhat differently from how it is conventionally understood. Instead of denoting a political philosophy, it is here understood as a particular (and fairly recent) 'art of government' (ibid., p. 210) and hence as a particular governmentality that makes use of a range of different techniques, including biopower over populations and the governing of the self (Lemke, 2002). It governs characteristically through using the subjects' 'capacities of acting and thinking' to secure the ends of government (Dean, 1999, p. 15), and it can therefore be said to govern through freedom (ibid.; Rose, 1996, p. 54).

These theoretical tools are used to examine how the problem of the relationship between the population and (water) resources is understood in discourses of sustainable development and water governance, and how these understandings are put to work to produce (new) ways of governing. Being especially concerned with how different populations – and lives – are understood and distinguished in sustainable development discourse, as well as inclusion/exclusion mechanisms in contemporary neoliberal governance, this book emphasises the drawing of lines between different forms of lives, both human and non-human. This focus demands that, while the book takes as its point of departure a Foucauldian understanding of governmentality and biopolitics, it draws on other biopolitical scholarship, such as Georgio Agamben's (1998) and Cary Wolfe's (2012) theories; scholarship which has placed a much greater emphasis on the ways in which biopolitical mechanisms make distinctions between different lives.

As we will see in the next chapter, such a distinction between different forms of lives, understood in relation to different (human) populations, has been a characteristic of population and environmental discourses, although in different shapes, since Malthus (1798). The curious fact that Foucault failed to engage properly with Malthusian logic in liberal thought (see Dean, 2015; Tellmann, 2013), serves here as an explanation of why Foucault's theories do not fully help us problematize the imaginaries of the way that different populations are understood in relation to resource use and futurity. In this book, this lacuna is also approached as a link between the applicability of Foucault's and Agamben's theories in the context of resource governance. While Foucault places the problem of the population at the centre of liberal thought, Agamben's theories aid the process of problematizing the distinctions made through how the different lives of those populations are understood.

The awareness of this lacuna in the way that biopolitics is comprehended within a Foucauldian framework, also draws attention to another important feature that has been lacking in much biopolitical scholarship, namely, how colonialism and colonial techniques of power are constitutive parts of biopolitics (see, for example, Alt, 2016; Venn, 2006). The way different populations were originally understood in the Malthusian populations thesis, was based on a colonial imaginary of the difference between civilized and savage life, which calls for a focus on the roles that these colonial imaginaries play with regard to resource and environmental policies.

In this book, these theories are applied in order to analyse the biopolitical aspects of global water governance. The concept of 'governmentality' is applied to provide a framework for analysing particular governing rationalities, ways of thinking, and practical techniques of rule through which water use and water users are governed. In this book these assemblages are termed 'hydromentalities'.

However, outcomes of governing are uncertain. Water users are not just passive subjects who are successfully governed to behave in certain ways; they are agents who actively navigate their lives. Therefore, when attempting to understand government in water management, and its *productive effects*, we need to emphasise the role that such governing has in terms of how water users themselves relate it to their lives and lifestyles, and, furthermore, how they produce certain subjectivities – as well as resistance strategies – in relation to them. Chapter 5 examines these effects in the local context of eThekwini through applying a narrative method.

In sum, this book engages with the governing logics of sustainable development in general, and water governance in particular, as well as their biopolitical effects in a local context. This means that this book is preoccupied with how governing strategies try to pin down the relationship between (different) populations and (water) resources, which paints a picture of a biopolitics that is premised on making distinctions between different forms of lives, both human and non-human. Furthermore, it shows how such exclusionary ways of governing resonate with water users' stories as some of the narrators construct themselves as something in between human and animal, as an excluded form of life, while others produce themselves as belonging to a privileged and 'qualified' form of life, to prove (bio)political points in relation to water.

There are also those who are left out of such strategies of governing more fundamentally, such as the stateless, migrants, and populations in failed states, in war zones, or suffering from famine. This aspect of a kind of a biopolitics of neglect and abandonment could fruitfully be analysed from an Agambenite perspective of biopolitics, which is acknowledged in this book, although its focus lies elsewhere. Another research theme that is worth studying within a biopolitical frame but which is not included here, is a systematic analysis of how the quality of water delivered to different areas and populations differs and what this means in relation to health and longevity. However, a detailed outline of what *does* feature in the book follows below.

Outline of the book

Chapter 2, entitled *(Green) governmentality and biopolitics*, explores the theoretical framing of the book. It outlines what a governmentality and biopolitical perspective entails for the study of sustainable development and resource governance, and how we can conceptualize both (neo)liberalism and the workings of power within a (green) governmentality framework

It also presents the features of contemporary biopolitics as part of a neoliberal governmentality, and it explains the basic concepts of biopolitics and governmentality, of how (neo)liberalism and power are approached and understood. A particular focus has been placed on the concept of biopolitics and how it has been developed by philosophers and political theorists such as Agamben, Esposito, and Wolfe. Central in relation to these theorists is the prominence given to the distinctions made between different forms of life, both human and non-human.

The chapter develops a green governmentality approach to environmental and resource governance by tracing the meaning of populations and scarcity in liberal economic thought. It also analyses the production of meaning around notions such as limits, scarcity, basic needs, and resilience in sustainable development discourse, and how these concepts relate to different forms of life and the hierarchization between them. In addition, the chapter stresses governing as uncertain and as something which therefore calls for empirical engagement in relation to how the governing logics and practices play out in different contexts and locations.

The framework of Chapter 2 provides theoretical insights into how neoliberal techniques of governing use the agency of free subjects in order to govern. Processes of responsibilization, together with the moralism argued to be a central component of discourses around sustainable development and care for the environment, are related to how subjects work on themselves through techniques of the self.

Chapter 3, *Global water governance: scarcity, populations, distinctions*, analyses global water governance (GWG) and draws the contours of GWG in terms of its emergence, its different actors, and its ways of working. The Integrated Water Resources Management (IWRM) approach and Water Security are discussed as central components of GWG, as is the role of water in the 2030 Agenda.

Chapter 3 outlines how we can understand GWG as *neoliberal* in a governmentality sense and analyses the governing rationales of GWG through a biopolitical reading. The role and meaning of scarcity within GWG are discussed, as are the three main rationales around the productive uses of water for (1) economic development, (2) basic human needs, and (3) ecosystems and how the notion of resilience has been considered within GWG. The chapter also describes central biopolitical aspects of such governance. One important conclusion is that poor people's relationship to water management, sustainability, and security has clear linkages with Malthus' stipulations. The chapter, furthermore, emphasises how a focus on meeting basic human needs through a human right to water can be problematized, because of how such an agenda involves a biopolitical separation between different populations. Moreover, the way that the ecological dimension of water management is understood in GWG makes explicit how present sustainable development discourses and GWG are based on a biopolitical distinction between humans and the rest of nature.

Another important point of this chapter is the governmentality understanding of the neoliberal characteristics of GWG, according to which, such governance not only includes market mechanisms as the ideal way of governing society, but is also premised on interventions in conventional economic models, such as the inclusion of ecosystem services and the adoption of the (human) right to water.

Chapter 4, *Hydromentalities*, engages with that concept as defined in this book (partly building on Mayhew, 2004 and Linton, 2010), which are assemblages of governing rationalities, techniques of rule and ways of thinking about, and defining, water. Hydromentalities involve particular ways of understanding what water is, the challenges involved in its allocation, and the proper way to manage and govern water use as well as water users. Based on this concept, the chapter discusses differences in the rationalities, mentalities, and techniques of water governance in relation to recent shifts, both historically and around the world. Thus, it outlines the shifts from the so-called hydraulic mission, to what this book terms neoliberal hydromentality, and provides insights into different water governance systems through analyses of different governmentalities of water in countries such as Ecuador (Boelens *et al.*, 2015), India (Babu, 2009; Birkenholtz, 2009), China (Rogers *et al.*, 2016), Niger (Hansson, 2014), and Indonesia (Kooy and Bakker, 2008a, 2008b).

Chapter 4 also outlines the particular hydromentality of eThekwini municipality in South Africa, one of the most contested sites when it comes to assessing progress in the implementation of South African water regulations. eThekwini is characterized by the notion of water as a right and a commodity, and by its practice of making a distinction between water users through the application of different technological artefacts and service levels for different individuals and populations. Another characteristic is the use of consultation, communication, and education to promote a particular way of understanding water and the challenges involved in its management.

The purpose of this book, however, is not only to scrutinize governing structures per se but also to study *what water is and what water does* for the water users. In Chapter 5, *Understanding biopolitical 'effects' locally: life(styles), subjectivities, and moral obligations*, such effects are studied through water users' narratives collected in eThekwini municipality.

Chapter 5 thus offers a methodological approach to studying the biopolitics of water in a particular location through the combination of biopolitical theory and narrative methods. Narrative methods were employed to produce text about how water users make sense of their (water) situation and how they convey a sense of self (Elliott, 2005, p. 126; Hinchman and Hinchman, 1997, p. xvi). When analysing water users' stories from a biopolitical perspective, attention is drawn to the ways in which water and water service delivery are tied to people's identities, and how water governance works to produce subjectivities and notions of different lifestyles, as well as notions of sustainability and responsibility. The analysis draws not only on Foucauldian scholarship but also on ideas of (ambivalent) distinctions made between different forms of lives as a central feature of modern biopolitics (as theorized by Agamben, 1998).

What we learn from the water users' narratives is that, first, water is not only important for survival, immediate needs, and health, but it is also vital for making people feel comfortable, safe, and good about themselves. In this sense, water makes people's lives 'liveable' as well as upholding survival. Second, the arrangement of water service delivery matters to people's lives to the extent that it affects the conditions for people's livelihoods and ways of life. Water governance is biopolitically performative through the water users' notions of Self and Other and distinctions between different forms of lives and lifestyles. In these processes, ambivalences between *zoë*, referring to 'the simple fact of living common to all living beings' (Agamben, 1998, p. 1), and *bios*, that is, 'the form or way of living proper to an individual or a group' (ibid.), are (re)produced. Third, technologies of government in water service delivery shape the way that certain subjectivities, as well as notions of ethical selves and an ethical allocation of water (both corresponding to and contrasting with governing logics), are formed, especially through particular notions of sustainability and responsibility.

In sum, the chapter contends that the biopolitics of eThekwini exacerbate the differences between individuals and populations, and that the way that (ir)responsible and (un)sustainable subjects are constructed is highly related to issues of class, race, and gender.

The concluding chapter of this book, Chapter 6 *The biopolitics of water*, suggests that the biopolitical framework outlined in this book is valuable for understanding the power and politics of water and water governance. It argues that a biopolitical perspective on water governance is useful for problematizing the population–resource nexus and how scarcity functions as a regulatory device in relation to different populations. Such an understanding of the biopolitical aspects of water governance makes explicit how different populations are viewed in relation to futurity and their chances of becoming 'sustainable' and, furthermore, how this biopolitical dispositif works to produce, and make appear natural, hierarchizations of different forms of lives, both human and non-human. Ultimately, these distinctions and hierarchizations are part of a governing assemblage that contributes to sustaining the divide between rich and poor populations and keeping intact the privileges and lifestyles of the middle classes of the Global North.

While bearing in mind the uncertainty of governing and the distinct characters of different contexts, the final chapter argues that an engagement with the notion of hydromentalities can be a fruitful way of scrutinizing the different shapes that water governance takes in different places and settings. In the course of which we should be wary of approaching governmentalities and hydromentalities as realized projects, but should treat their successes and effects as an empirical question.

In relation to the discussion on the right to *basic* water, which is a central debate in water governance literature and a crucial part of the biopolitical aspect of water governance, Chapter 6 argues that we should not be concerned (only) with finding appropriate figures to account for the amount of basic water, or how to design welfare-maximizing tariffs. Rather, the important discussion that emerges from the analysis in this book, is that we should be questioning the drawing of a line between *basic water* and water used for other purposes, as current practice – along with a commodification above the threshold of survival – risks further entrenching distinctions between different forms of life. Furthermore, and relatedly, whether we draw this line or not, it is not possible to separate the biological work that water does in keeping bodies alive, from the political work it does in reaffirming people's sense of self and their place in society.

Finally, the concluding chapter contends that in the search for a more just way of allocating water, we need to reverse the stipulations made by Malthus and place responsibility instead on affluent individuals and populations, who are the ones who truly think and act without a sense of futurity.

Notes

1 IWRM is commonly defined as 'the co-ordinated development and management of water, land and related resources, in order to maximize the resultant economic and social welfare in an equitable manner without compromising the sustainability of vital ecosystems' (GWP, 2000, p. 22).

2 Initially (in the 1990s), the notion of a 'water war' was first centred around armed conflicts between states (see Starr, 1991). This was a consequence of the fact that the framing of the water crisis led to a subsequent framing of water as a security issue. This issue later became part of the, so-called, broadening of the security concept through the notion of human security (see Brauch, 2008). Subsequently, however, water wars have come to mean something other than wars over a water resource between states or regions. Shiva (2002) uses the term to describe linguistic struggles over the representation and definitions of water. Today, it is used to depict local struggles over water access.

References

Agamben, G. (1998) *Homo Sacer: Sovereign power and Bare Life*. Stanford, CA: Stanford University Press.

Ahlers, R. and Zwarteveen, M. (2009) 'The water question in feminism: water control and gender inequalities in a neoliberal-era', *Gender, Place and Culture: A Journal of Feminist Geography*, 16(4), 409–426.

Alt, S. (2016) 'Beyond the Biopolitics of Development: Being, Politics and Worlds', PhD thesis, University of Lapland.

Appadurai, A. (2001) 'Deep democracy: urban governmentality and the horizon of politics', *Environment and Urbanization*, 13(2), 23–43.

Babu, A. (2009) 'Governmentality, Active Citizenship and Marginalisation: The Case of Rural Drinking Water supply in Kerala, India', *Asian Social Science*, 5(11), 89–98.

Bakker, K. (2005) 'Neoliberalizing nature? Market environmentalism in water supply in England and Wales', *Annals of the Association of American Geographers*, 95(3), 542–565.

Bakker, K. (2007) 'The "Commons" Versus the "Commodity": Alter-globalization, Anti-privatization and the Human Right to Water in the Global South', *Antipode* 39(3), 430–455.

Bakker, K. (2010) *Privatizing Water: Governance Failure and the World's Urban Water crisis*. Ithaca, NY: Cornell University Press.

Bakker, K. (2012) 'Water: Political, Biopolitical, Material', *Social Studies of Science*, 42(4), 616–623.

Bakker, K. (2013) 'Constructing "Public" Water: The World Bank, Urban Water Supply, and the Biopolitics of Development', *Environment and Planning D: Society and Space*, 31(2), 280–300.

Barker, R. and Molle, F. (2004) *Evolution of Irrigation in South and Southeast Asia. Colombo and Sri Lanka: Comprehensive Assessment Secretariat.* (Comprehensive Assessment Research Report 5).

Barlow, M. and Clarke, T. (2003) *Blue Gold: The Fight to Stop the Corporate Theft of the World's Water.* New York, NY: Stoddart.

Bashford, A. (2006) 'Global biopolitics and the history of world health', *History of the Human Sciences*, 19, 67–88.

Birkenholtz, T. (2008) 'Contesting Expertise: The politics of environmental knowledge in northern Indian Groundwater practices', *Geoforum*, 39, 466–482.

Birkenholtz, T. (2009) 'Groundwater Governmentality: hegemony and technologies of resistance in Rajasthan's (India) groundwater governance', *The Geographical Journal*, 175(3), 208–220.

Birkenholtz, T. (2010) ' "Full-cost recovery": producing differentiated water collection practises and responses to centralized water networks in Jaipur', *Environment & Planning A*, 42: 2238–2253.

Birkenholtz, T. (2013) ' "On the network, off the map": developing intervillage and intragender differentiation in rural water supply', *Environment and Planning D: Society and Space*, 31(2), 354–371.

Boelens, R., Hoogesteger, J. and Baud, M. (2015) 'Water reform governmentality in Ecuador: Neoliberalism, centralization, and the restraining of polycentric authority and community rule-making', *Geoforum*, 64, 281–291.

Brauch, H. G. (2008) 'Conceptualising the environmental dimension of human security in the UN', *International Social Science Journal*, 59, 19–48.

Braun, B. and Whatmore, S. (eds.) (2010) *Political Matter: Technoscience, Democracy, and Public Life.* Minneapolis, MN: University of Minnesota Press.

Butler, J. (2004) *Undoing Gender.* London, UK: Routledge.

Castro, J. E. (2008) 'Water Struggles, Citizenship and Governance in Latin America', *Society for International Development*, 51, 72–76

Chatterjee, P. (2004) *Politics of the Governed: Reflections on Popular Politics in Most of the World.* New York, NY: Columbia University Press.

Conca, K. (2006) *Governing Water.* Cambridge, MA: MIT Press.

Davies, M. I. J. (2009) 'Wittfogel's dilemma: heterarchy and ethnographic approaches to irrigation management in Eastern Africa and Mesopotamia', *World Archaeology*, 41: 16–35

Dean, M. (1999) *Governmentality: Power and Rule in Modern Society.* Los Angeles, CA: Sage.

Dean, M. (2015) 'The Malthus Effect: population and the liberal government of life', *Economy and Society*, 44(1), 18–39.

Ekers, M. and Loftus, A. (2008) 'The power of water: developing dialogues between Foucault and Gramsci', *Environment and Planning D: Society and Space*, 26(4), 698–718.

Elliott, J. (2005) *Using Narrative in Social Research: Qualitative and Quantitative Approaches.* Los Angeles, CA: Sage.

Foucault, M. (1998) [1976] *The Will to Knowledge: History of Sexuality I.* Harmondsworth, UK: Penguin.

Foucault, M. (2003) *"Society Must Be Defended", Lectures at the Collège de France 1975–1976.* New York, NY: Picador.

Gandy, M. (2004) 'Rethinking urban metabolism: Water, space and the modern city', *City*, 8(3), 363–379.

Gandy, M. (2006) 'Zones of indistinction: some thoughts on the bio-politics of urban space', *Cultural Geographies*, 13(4), 1–20.

Gandy, M. (2014) *The fabric of space: water, modernity, and the urban imagination*. Cambridge, MA: The MIT Press.

Gibbs, L. M. (2006) 'Valuing water: Variability and the Lake Eyre Basin, Central Australia', *Australian Geographer*, 37(1), 73–85.

Gibbs, L. M. (2009) 'Water places: cultural, social and more-than-human geographies of nature', *Scottish Geographical Journal*, 125(3–4), 361–369.

Gleick, P. H. (ed.) (1993) *Water in Crisis: A Guide to the World's Freshwater Resources*. New York, NY: Oxford University Press.

Goubert, J. P. (1989) *The Conquest of Water: The Advent of Health in the Industrial Age*. Cambridge, UK: Polity Press

GWP (Global Water Partnership). (2000) *TAC Background Paper No. 4: Integrated Water Resources Management*. Denmark: Global Water Partnership.

Hansson, S. (2014) 'Dependence and Autonomy: Conditions of Possibility for Nigerien Ownership in the Water Sector', *Journal of Intervention and Statebuilding*, 9(2), 256–277.

Harris, L. (2008) 'Water Rich, Resource Poor: Intersections of Gender, Poverty, and Vulnerability in Newly Irrigated Areas of Southeastern Turkey', *World Development*, 36(12), 2643–2662.

Harris, L. (2009) 'Gender and emergent water governance: comparative overview of neoliberalized natures and gender dimensions of privatization, devolution and marketization', *Gender, Place and Culture*, 16(4), 387–408.

Hellberg, S. (2014) 'Water, life and politics: Exploring the contested case of eThekwini municipality through a governmentality lens', *Geoforum*, 56, 226–326.

Hellberg, S. (2017) 'Water for survival, water for pleasure – A biopolitical perspective on the social sustainability of the basic water agenda', *Water Alternatives*, 10(1), 65–80.

Heynen, N., Kaika, M. and Swyngedouw, E. (2005) 'Urban Political Ecology – Politicising the Production of Urban Natures', in Heynen, N., Kaika, M. and Swyngedouw, E. (eds.) *In the Nature of Cities – Urban Political Ecology and The Politics of Urban Metabolism*. London, UK: Routledge.

Hinchman, L. P. and Hinchman, D. K. (eds.) (1997) *Memory, Identity, Community: The Idea of Narrative in Human Sciences*. New York, NY: State University of New York Press.

Johnston, B. R. (2003) 'The Political Ecology of Water: An Introduction', *Capitalism, Nature, Socialism*, 14(3): 73–90

Joyce, P. (2003) *The Rule of Freedom: Liberalism and the Modern City*. London and New York: Verso.

Kabeer, N. (2002) 'Citizenship, affiliation and exclusion: perspectives from the south', *IDS Bulletin*, 33, 1–15.

Kaika, M. (2005) *City of Flows: Modernity, Nature and the City*. London: Routledge.

Kooy, M. and Bakker, K. (2008a) 'Splintered networks: Urban water governance in Jakarta', *Geoforum*, 39(6), 1843–1858.

Kooy, M. and Bakker, K. (2008b) 'Technologies of Government: Constituting Subjectivities, Spaces, and Infrastructures in Colonial and Contemporary Jakarta', *International Journal of Urban and Regional Research*, 32(2), 375–391.

Legg, S. (2006) 'Governmentality, congestion, and calculation in colonial Delhi', *Social and Cultural Geography*, 7, 709–728.

Lemke, T. (2002) 'Foucault, Governmentality, and Critique', *Rethinking Marxism*, 14(3).

Lemke, T. (2007) 'An indigestible meal? Foucault, governmentality and state theory', *Distinktion: Journal of Social Theory*, 8 (2), 43–64.

Li, T. (2007) *The Will to Improve: Governmentality, Development and the Practice of Politics*. Durham, NC: Duke University Press.

Linton, J. (2010) *What is Water? The History of a Modern Abstraction*. Vancouver, BC: UBS Press.

Linton, J. and Budds, J. (2014) 'The hydrosocial cycle: Defining and mobilizing a relational-dialectical approach to water', *Geoforum*, 57, 170–180.

Loftus, A. (2005a) 'A Political Ecology of Water Struggles in Durban, South Africa', PhD thesis, University of Oxford.

Loftus, A. (2005b) '"Free Water" as Commodity: The Paradoxes of Durban's Water Service Transformations', in McDonald, D. A. and Ruiters, G. (eds.) *The Age of Commodity: Water Privatization in Southern Africa*. London, UK: Earthscan.

Loftus, A. (2006) 'The metabolic processes of capital accumulation in Durban's waterscape', in Heynen, N., Kaika, M., and Swyngedouw, E. (eds.). *In the Nature of Cities: Urban political ecology and the politics of urban metabolism*. London, UK: Routledge.

Loftus, A. (2007) 'Working the Socio-Natural Relations of the Urban Waterscape in South Africa', *International Journal and Regional Research*, 31(1), 41–59.

Loftus, A. (2009) 'Rethinking Political Ecologies of Water', *Third World Quarterly*, 30(5), 953–968

Loftus, A. (2015) 'Water (in)security: securing the right to water', *The Geographical Journal*, 181(4), 350–356.

Malthus, T. (1798) *An Essay on the Principle of Population*. London, UK: Dent & Sons.

Mamdani, M. (1996) *Citizen and Subject: Contemporary Africa and the Legacy of Late Colonialism*. Princeton, NJ: Princeton University Press.

Mayhew, S. (2004) 'Governmentality', in *A Dictionary of Geography*. Oxford: Oxford University Press.

McDonald, D. A. and Pape, J. (2002) *Cost Recovery and the Crisis of Service Delivery in South Africa*. Cape Town, RSA: Zed Press and HCRS Publishers.

McDonald, D. A. and Ruiters, G. (eds.) (2005) *The Age of Commodity: Water Privatization in Southern Africa*. London, UK: Earthscan.

Molle, F., Mollinga P. P., and Wester, P. (2009) 'Hydraulic Bureaucracies and the Hydraulic Mission: Flows of Water, Flows of Power', *Water Alternatives*, 2(3), 328-349.

Oliviera, O. (2004) *¡Cochabamba!: Water War In Bolivia*. New York, NY: South End Press.

O'Reilly, K. (2006) '"Traditional" Women, "Modern": Linking gender and commodification in Rajasthan, India', *Geoforum*, 37(6), 958–972.

Rodina, L. and Harris, L. M (2016) 'Water services, lived citizenship, and notions of the state in marginalised urban spaces: The case of Khayelitsha, South Africa', *Water Alternatives* 9(2), 336–355.

Rogers, S., Barnett, J., Webber, M., Finlayson, B., and Wang, M. (2016) 'Governmentality and the Conduct of Water: China's South–North Water Transfer Project', *Transaction of the Institute of British Geographers*. Doi: 10.1111/tran.12141

Rose, N. (1996) 'Governing "advanced" liberal democracies', in Barry, A., Osborne, T., and Rose, N. (eds.) *Foucault and Political Reason. Liberalism, neo-liberalism and rationalities of government*. Chicago, IL: The University of Chicago Press.

RSA (Republic of South Africa). (1996) *Constitution of the Republic of South Africa*, Act 108 of 1996. Pretoria: Government Printer.

Shiklomanov, I. A. (1993) 'World Fresh Water Resources', in Gleick, P. H., *Water in Crisis: A Guide to the World's Freshwater Resources*. New York: Oxford University Press.

Shiva, V. (2002) *Water Wars: Privatization, Pollution and Profit*. Cambridge, MA: Sound End Press.

Shove, E. (2003) *Comfort, Cleanliness and Convenience: The Social Organization of Normality*. Oxford and New York: Berg.

Sofoulis, Z. (2005) 'Big Water, Everyday Water: A Sociotechnical Perspective', *Continuum: Journal of Media and Cultural Studies*, 19(4), 445–463.

Strang, V. (2004) *The Meaning of Water*. Oxford, UK: Berg Publishers.

Strang, V. (2009) *Gardening the World: Agency, Identity, and the Ownership of Water*. New York and Oxford: Berghahn Books.

Starr, J. (1991) 'Water Wars', *Foreign Policy*, 82, 17–36.

Sultana, F. (2009) 'Fluid lives: subjectivities, gender and water in rural Bangladesh', *Gender, Place and Culture*, 16(4), 427–444.

Sultana, F. (2011) 'Suffering for Water, Suffering from Water: Emotional Geographies of Resource Access, Control and Conflict', *Geoforum*, 42(2), 163–172.

Sultana, F. (2013) 'Water technology, and development: transformations of development technonatures in changing waterscapes', *Environment and Planning D: Society and Space*, (31): 337–353.

Sultana, F. and Loftus, A. (2012) *The Right to Water: Politics, Governance and Social Struggles*. London, UK: Routledge.

Swyngedouw, E. (1999) 'Modernity and Hybridity: Nature, Regeneracionismo, and the Production of the Spanish Waterscape, 1890–1930', *Annals of the Association of American Geographers*, 89(3), 443–465.

Swyngedouw, E. (2004) *Social Power and the Urbanization of Water: Flows of Power*. Oxford, UK: Oxford University Press.

Swyngedouw, E. (2005) 'Dispossessing H_2O: The contested terrain of water privatization', *Capitalism Nature Socialism*, 16(1), 81–98.

Swyngedouw, E. (2006) 'Circulations and Metabolisms: (Hybrid) Natures and (Cyborg) Cities', *Science as Culture*, 15(2), 105–122.

Swyngedouw, E. (2009) The Political Economy and Political Ecology of the Hydro-Social Cycle. *Journal of Contemporary Water Research and Education*, 142(1), 56–60.

Swyngedouw, E. (2015) *Liquid Power: Contested Hydro-Modernities in Twentieth-Century Spain*. Cambridge, Massachusetts, London, England: MIT Press.

Tellmann, U. (2013) 'Catastrophic Populations and the Fear of the Future: Malthus and the Genealogy of Liberal Economy', *Theory, Culture and Society*, 30(2), 135–155.

Turton, A. R. (2002) 'Hydropolitics: The concept and its limitations', in Turton, A. and Henwood, R. (eds.) *Hydropolitics in the developing world: A Southern African Perspective*. Pretoria, RSA: African Water Research Unit

Venn, C. (2006) *The Postcolonial Challenge: Towards Alternative Worlds*. London, UK: Sage.

von Schnitzler, A. (2008) 'Citizenship Prepaid: Water, Calculability, and Techno-Politics in South Africa', *Journal of Southern Africa Studies*, 34(4).

Ward, L. (2013) 'Eco-governmentality revisited: Mapping divergent subjectivities among Integrated Water Resources Management experts in Paraguay', *Geoforum*, 46, 91–102.

Waterbury, J. (1979) *Hydropolitics of the Nile Valley*. New York: Syracuse University Press.

Wittfogel, K. (1957) *Oriental Despotism: A Comparative Study of Total Power*. New York, NY: Random House.

Wolfe, C. (2012) *Before the law: humans and other animals in a biopolitical frame*. Chicago, IL: University of Chicago Press.

WWDR (World Water Development Report). (2006) *Water – a shared responsibility*. UNESCO, http://unesdoc.unesco.org/images/0014/001454/145405E.pdf [accessed 8 October 2017].

2 (Green) governmentality and biopolitics

Introduction

Chapter 1 ended with a presentation of water research that uses Foucauldian theory to address power relations and modes of governing in water politics and management. It also gave a short presentation of what a governmentality/biopolitical perspective involves. Chapter 2 takes a step back to describe the theoretical foundations of this book in more detail.

In order to study the power and politics of water, the theoretical framework is constituted around Foucauldian notions of power and the different techniques it uses at it takes shape in neoliberal governmentality. Biopower and biopolitics are especially important terms for capturing the ways that water governance works in a sustainable development regime. Foucauldian notions are complemented by insights from other political theorists who have developed Foucault's theories, such as Giorgio Agamben, Roberto Esposito, and Cary Wolfe.

This chapter describes how insights in governmentality and biopolitics literature have been understood within the more specific field of *green governmentality*. It discusses what a green governmentality perspective entails for the study of resource governance, and how we can use these insights for a further investigation of the politics of sustainable development and resource governance.

The first section introduces basic concepts and points of departure in a Foucauldian analysis, including the concepts of governmentality and biopolitics and how (neo)liberalism and power are approached and understood. The concept of biopolitics is discussed in relation to contributions from theorists who have developed (and contrasted) Foucauldian notions of biopolitics.

The second section then outlines how such understandings have been applied in studies of the environment and presents two central themes of the book: the uncertainty of governing and the making of (green) subjectivities in relations of power. The third section develops a green governmentality approach to environmental and resource governance by tracing the meaning of populations and scarcity in liberal economic thought. It also analyses the production of meaning around notions such as limits, scarcity, basic needs, and resilience in sustainable development discourse, and how these concepts relate to different forms of life, both human and non-human, and the hierarchization between them.

A governmentality/biopolitics framework

This book applies a governmentality/biopolitics perspective to the study of water governance. Governmentality theory addresses the how-question of governing/government. The term governmentality is understood as 'the organized practices (mentalities, rationalities, and techniques) through which subjects are governed' (Mayhew, 2004), and it describes how government works in practice through a variety of techniques, including biopower over populations and the governing of the self (Lemke, 2002). In neoliberal governmentality, biopolitics, which is a specific form of politics directed at the life conditions of populations, constitutes one aspect, or rather a 'necessary condition', of modern neoliberal governmentality (Dean, 1999, p. 113). This chapter outlines central features of a governmentality/biopolitics perspective and explains how such a framework is productive for the task of analysing power relations in water governance, starting with a discussion of basic concepts and understandings.

(Neo)liberalism

One basic point it is important to make immediately is the way that (neo)liberalism is conceptualized from a Foucauldian approach. From such a perspective, liberalism is seen as a practical 'art of government' rather than as a political philosophy (Dean, 1999, p. 210). Liberalism should not be understood as the product of a system that is 'free' but that, for liberal freedoms to come into being – such as freedom of the market and freedom of expression – liberalism 'must produce freedom, but this very act entails the establishment of limitations, controls and forms of coercion, and obligations relying on threats etcetera' (Foucault, 2008, p. 64). Moreover, in a framework focusing on neoliberal governmentality, freedom is more than anything a governing technique, one which uses subjects' freedom, that is, their 'capacities of acting and thinking', to secure the government's objectives (Dean, 1999, p. 15).

Foucault's theories help us grasp the difference between liberalism and neoliberalism. While the concern in liberalism was to 'cut out or contrive a free space of the market within an already given political society', neoliberalism is concerned with the problem of making the principles of market economy the model for 'the overall exercise of political power' (Foucault, 2008, p. 131). In this process, the market becomes 'that which reveals something like the truth' (ibid., p. 32), a site of veridiction, in relation to which governmental practices are assessed. Thus, according to

Foucault, we can understand the objective of neoliberalism as the attainment of 'a regulation of society by the market' (ibid.). According to this perspective, constant interventions are required to make society function in accordance with the market as the ideal model for its organisation. Such a perception differs substantially from an understanding of liberalism as involving less governmental intervention. On the contrary, Foucault argues that neoliberalism is to be seen as an art of government that requires government:

> no less dense, frequent, active, and continuous than in any other system.... Basically, it has to intervene in society so that competitive mechanisms can play a regulatory role at every moment and every point in society and by intervening in this way its objective will become possible, that is to say, a general regulation of society by the market.
>
> (Ibid., p. 145)

Techniques of power

Another basic point of departure for a Foucauldian approach is the way that power is understood and the different modalities through which it functions. In a Foucauldian framework, power is seen as productive, a position that has become commonplace in most contemporary discussions of power. According to Foucault, 'relations of power ... have a directly productive role, whenever they come into play' (1998, p. 94):

> We must cease once and for all to describe the effects of power in negative terms: 'it 'excludes', it 'represses', it 'censors', it 'abstracts', it 'masks', it 'conceals'. In fact, power produces; it produces reality; it produces domains of objects and rituals of truth. The individual and the knowledge that may be gained of him belong to this production.
>
> (Foucault, 1977, p. 194)

In Foucault's writings, power cannot be possessed by individuals; it cannot be 'acquired, seized, or shared' (1998, p. 94) as it is 'employed and exercised through a net-like organisation' (Foucault, 1980, p. 98). Based on such an understanding, power does not involve an 'all-encompassing opposition between rulers and ruled' (Foucault, 1998, p. 94) in the sense of one actor fully dominating the other. On the contrary, freedom is an essential component of, and a prerequisite for, power relations. Foucault states that 'in order for power relations to come into play, there must be at least a certain degree of freedom on both sides' (1997, p. 292). This means that the subjects over (or through) whom power is exercised, 'are faced with a field of possibilities in which several kinds of conduct, several ways of reacting and modes of behaviour are available' (Foucault, 2000, p. 342).[1]

Such a way of comprehending how power works through subjects is of fundamental importance when understanding water governance from a Foucauldian perspective, since it acknowledges and places an emphasis on the different possibilities of acting and thinking (differently) in relation to governing logics of water management. However, since the subjects' freedom and agency are made use of in governing strategies, studying this involves methodological challenges because of the impossibility of distinguishing between a power that comes from 'outside' and the subject's 'own' power (Butler, 1997, p. 15), a challenge returned to in Chapter 5. The way that subjects form, or conduct themselves, Foucault termed 'technologies of the self', which:

> permit individuals to effect by their own means or with the help of others a certain number of operations on their own bodies and souls, thoughts, conduct, and way of being, so as to transform themselves in order to attain a certain state of happiness, purity, wisdom, perfection, or immortality.
>
> (Foucault, 1988, p. 18)

Hence, technologies of the self are to be understood as ways for individuals to set rules for conduct and to transform themselves and their lives in relation to certain aesthetic values and criteria and to shape themselves as ethical subjects (ibid., pp. 10–11, 13). For individuals, this involves a problematization of 'what they are, what they do, and the world in which they live', something Foucault argues is central to our (modern, Western) societies (ibid., p. 10). Thus, technologies of the self are ways for subjects to form themselves within relations of power (which are present in all social beings and in relation to governing strategies).

According to a Foucauldian understanding, power in modern governmentality is exercised through different techniques: sovereign power, disciplinary power, and biopower.[2] Foucault relates *sovereign power* to the authority of the sovereign state and the obedience to its laws and central figure(s). Such an understanding of power involves the assumption that power has a source or a central point (Foucault, 2003, p. 265). His work on *disciplinary power* should be seen as a reaction to those approaches to power that take 'the law as the basic manifestation of power' (ibid.). In contrast to sovereign power – which seeks revenge for breaking the law or disobeying the king – disciplinary techniques aim to correct deviant behaviour through the spatial distribution of bodies and through observation, examination, and surveillance (Foucault, 1977).[3] *Biopower*, on the other hand, turns away from the subject/individual and works *at the level of the population*. It is this technique of power, and how its emergence changed the government of modern societies, that is discussed below. It is important to note that the advent of biopower does not mean that the other techniques of power have disappeared. Rather, those modalities of power are put to work simultaneously in biopolitical strategies of governing.

Biopower and biopolitics

Biopower, Foucault argued, developed in the second half of the eighteenth century. It was then that life made its entry 'into history' (1998, p. 141). Foucault explains:

> For the first time in history, no doubt, biological existence was reflected in political existence; the fact of living was no longer an inaccessible substrate that only emerged from time to time, amid the randomness of death and its fatality; part of it passed into knowledge's field of control and power's sphere of intervention.
>
> (Ibid., 1998, p. 142)

From the advent of biopower, life became a target of explicit calculations, and assemblages of power-knowledge undertook to transform it (ibid., p. 143). The mechanisms introduced by this technology of power are different from those of disciplinary power. For example, they include statistics and models for making measurements and predictions (Foucault, 2003, p. 246), to establish an average at population level and to compensate for variations (ibid.).

As a consequence of his theoretical development of disciplinary power and biopower, Foucault argued that we should 'cut off the head of the king' (see, for example, Foucault, 1998, p. 89). By this he meant that analysing power relations in their concrete configurations, requires 'abandon[ing] the juridical model of sovereignty' (Foucault, 2003, p. 265). This does not, however, mean that sovereign power is disappearing as a result of the emergence of other techniques of power (disciplinary power and biopower).[4,5] Rather, it is a shift of focus, from the law and the juridical system to that of government and regulation:

> I do not mean to say that the law fades into the background or that the institutions of justice tend to disappear but rather that the law operates more and more as a norm, and that the judicial institution is increasingly incorporated into a continuum of apparatuses (medical, administrative and so on) whose functions are for the most part regulatory. A normalizing society is the historical outcome of a technology of power centred on life.
>
> (Foucault, 1998, p. 144)

Thus, the development of biopower, rather than replacing other techniques of power, transforms them:

> This technology of power does not exclude the former, does not exclude disciplinary technology, but it does dovetail into it, integrate it, modify it to some extent, and above all, use it by sort of infiltrating it, embedding itself in existing disciplinary techniques. This new technique does

> not simply do away with the disciplinary technique, because it exists at a different level, on a different scale, and because it has a different bearing area, and makes use of very different instruments.
>
> (Foucault, 2003, p. 242)

Foucault states: '[a] power whose task it is to take charge of life needs continuous regulatory and corrective mechanisms' (Foucault, 1998, p. 144). Such a *regulatory* society can be understood as a form of government that attends far less to detail – letting things take their course instead (i.e. laissez-faire) – than in a disciplinary society (Foucault, 2007, p. 45). This idea is developed further on in the chapter in connection with the theme of scarcity, a central theme in relation to the politics of sustainable development and of water governance.

Again, it is important to note that 'freedom' in liberalism is made use of in order to govern 'at a distance' (Rose and Miller, 1992, p. 181). In a regulatory society, the subjects of governments are enjoined to be active in their own government and to work on themselves to become subjects/citizens whose interests correspond to those of society as a whole (Cruikshank, 1999, p. 40). Such insights are core to the literature on (neoliberal) governmentality. Below is a description of the development of the notion of biopolitics beyond Foucault.

Biopolitics and the separation of different forms of lives

It is beyond the scope of this book to describe, let alone settle, the debates regarding biopolitics. It is, however, important to introduce the main biopolitical strands of thought regarding how contemporary politics makes distinctions, or draws lines, between different forms of life, which is done below by focusing mainly on Agamben's contributions to biopolitical theory, while also acknowledging Roberto Esposito, Judith Butler, and Cary Wolfe as important thinkers within this debate (as well as implicitly, through the work of the above-mentioned theorists, Heidegger and Arendt).[6] Also by outlining some of the thoughts on how lines are drawn (or not) between humans and other animals, a theme that has been less explored in biopolitical theory (Wolfe, 2012).

Agamben is one of the most renowned scholars to have developed the notion of biopolitics beyond Foucault's understandings. In contrast to Foucault, Agamben argues that the power over life has been implicit in the whole of Western political thought (and is not a specific characteristic of modernity, as Foucault claims [see 1998, p. 143]).

According to Agamben, biopower is not distinct from sovereign power. Rather, the production of the biopolitical body is the 'original activity of sovereign power' (Agamben, 1998, p. 6). It functions so as to make distinctions between different forms of life (Edkins and Pin-Fat, 2004, p. 6). Central to Agamben's work is the distinction between *zoë*, referring to 'the simple fact of living common to all living beings' (Agamben, 1998, p. 1) and *bios*, meaning 'the form or way of living proper to an individual or a group' (ibid.), that is, the difference between biological life and the political life (or mere life and the good life [Plonowska Ziarek, 2012, p. 1]). Equally important is the notion of *bare life*.

In ancient Roman law, bare life was exemplified by *homo sacer* – a figure in Roman law who is banned on the basis of a crime and therefore can be killed (by anyone, with impunity), but whose death cannot be seen as a sacrifice because *homo sacer* is excluded from right and law. It is, hence, only through the act of keeping *homo sacer* outside the community of rights (Sylvester, 2006, p. 67) that he will get his role in the sovereign system. Or, in other words, it is only through the act of law that *homo sacer* becomes lawless and is included by exclusion.

In modern biopolitics, however, this ancient ban is replaced by a disjunctive inclusion of bare life 'into the political realm and the structure of citizenship' (Plonowska Ziarek, 2012, p. 2) and hence into the system of rights. This does not mean, however, that bare life becomes transformed into 'political existence' but, rather, that it is still the target of sovereign violence (ibid.). Bare life is excluded life. This exclusion is a performative act displaying the difference between bare life and political life, and through this difference the latter can understand itself as qualified (Wrangel, 2014, p. 5).

Yet again, there is a difference between how Agamben understands the politics of life and Foucauldian biopolitics. According to Mika Ojakangas (2005), the concept of life in Foucauldian biopolitics is 'univocal and immanent to itself'. Thus, in a Foucauldian understanding there is no need for, or possibility of, distinguishing between *bios* and *zoë*, between *forms of life* and *bare life*. Ojakangas writes: 'Instead of *homo sacer* ... the paradigmatic figure of the biopolitical society can be seen, for example, in the middle-class Swedish social democrat' (ibid., p. 1). The point here is that in a Foucauldian understanding, the focus of biopolitical strategies of governing are to *improve* and *optimize life* (within the limits of the political economy) and in which life always is political, while biopolitics in an Agambenite understanding revolve around the distinctions between life that is political and other forms of lives.

Agamben's theories are also important in terms of theorising the ambivalent relation to the notion of life that is symptomatic of liberal biopolitics. According to Agamben, modern biopolitics involves a political contradiction in the sense of not having 'succeeded in constructing the link between *zoë* and *bios*.... Bare life remains included in politics in the form of the exception, that is, as something that is included solely through exclusion' (1998, p. 11). One of the basic characteristics of modern biopolitics is therefore that it is 'in constant need to redefine the threshold in life that distinguishes and separates what is inside from what is outside' (ibid., p. 131).

Esposito has yet another take on the inclusion/exclusion dynamics of biopolitics. According to this Italian philosopher, the central feature of modern biopolitics is how it secures the community against threats through *immunization*. According to Esposito this is, 'a negative [form] of the protection of life' (2008b, p. 46, square brackets in original). He explains that immunization of the political body functions in the same way as vaccinating an individual body. 'It saves, insures and preserves the organism, either individual or collective' through introducing a fragment within that body 'from which it wants to protect itself by blocking and contracting natural development' (ibid., p. 46). In juridical terms, immunization means to be safeguarded and untouchable in relation to common law (2008a, p. 3). In both cases, being immune involves a situation in which someone is protected from a risk the entire community is exposed to. The important point here is that this means there is an opposition between the notions of *immunity* and *community*, since the process of immunization means that the immune person places him- or herself outside social circulation (ibid.).

Immunization is an exclusionary and violent process (Alt, 2016, p. 139). Not only does it involve a situation of 'a sort of cage' (Esposito 2008a, p. 7), where both freedom and an individual and collective existence outside of itself (and hence outside of the community) are lost. Immunization 'at high doses', according to Esposito, reduces 'every form of qualified life, to simple survival. The reduction of life to its bare biological layer, of *bios* to *zoe* [*sic*]' (ibid., italics in original). At its extreme, it also involves a process of destroying not only the external enemy but also its own body (ibid., p. 11).

Thus, Esposito diagnoses contemporary biopolitics within the paradigm of immunization, in which life is negatively protected in a system that does not tolerate other truths or logics, or even the idea of an outside (ibid.), to the point that it becomes violent and lethal. In the West, such a logic of 'the One-everything' (ibid.) is represented by a political monotheism guided by the principles of 'technological performance, the logic of making money, and total production' (ibid.).

Esposito's task is not only to describe an immunization paradigm, he also asks what can be done 'to break out of this logic of death' (ibid.). The answer is an 'affirmative biopolitics' that is based on the opposite of *immunitas*, namely, 'the open and plural form of *communitas*' (ibid., p. 13). In Esposito's thinking, such a community forms part of a de-individualized 'flesh of the world' that embraces not only difference but also risk and vulnerability (Watson, 2012).

As Janell Watson (2012) points out, such reasoning has clear links to Butler's work on precarity (2006, 2009), which embraces a vulnerability that can form a remedy for immunization and for the violence that follows from the national subject that tries to immunize itself against perceived threats. Such a shared vulnerability Butler terms *precariousness*, which is a corporeal vulnerability that all mortals share. This term should be understood in relation to, and is different from, *precarity*, which is the particular vulnerability that is imposed on the marginalised (Butler, 2009, p. 25; Watson, 2012). In Butler's writing, avowing precariousness – especially by the powerful, those who deny their precariousness – is a remedy against the immunization logic because it is a social condition (rather than an individual existential one) that can produce certain political demands and principles, such as non-military political solutions (Watson, 2012).

Both Esposito and Butler have been criticized for the way they propose vulnerability as a strategy for political resistance (Alt, 2016). Embracing vulnerability, the critics argue, is part of neoliberal governmentality and its way of legitimizing its power (Reid, 2011). Furthermore, a politics based on shared vulnerability, according to the critics, fails to account or offer remedies for 'unequally imposed precarity' (Watson, 2012).

This argument relates to another lacuna identified in the (early) biopolitics literature, namely, the failure to address imperialism in the emergence of biopower and coloniality in biopolitical strategies of governing (Alt, 2016). Being occupied with the questions of modernity in the development of theories of biopolitics, Foucault, for example, does not explore the ways in which colonialism and colonial techniques of power have been constitutive of modern forms of (bio)power and governmentality. Thinkers such as Gayatri Chakravorty Spivak (1999) and Couze Venn (2006) have brought attention to this lack (Alt, 2016, p. 66). In the context of this book and the question of distinctions made between different forms of lives, acknowledging the eurocentrism of early biopolitical thought calls for a focus on the role played by colonial logics in terms of approaching, defining, and neglecting different populations and individuals.

Another central but somewhat neglected aspect of biopolitics is the distinction made between humans and other animals (Stanescu, 2013; Wolfe, 2012). In terms of distinguishing between different forms of lives, James Stanescu argues that 'the line drawn between the human animal and other animals is one of the, if not the most, important division in the biopolitical terrain' (2013, p. 135). This division is effectively deconstructed by Wolfe, not only because the distinction between humans and other animals is blurred but also because animals (other than humans) differ from humans in different ways (2012, pp. 73–74).

Unlike Esposito's efforts to find a solution to the violence and exclusion that biopolitics entails through his affirmative biopolitics, Wolfe approaches biopolitics as a 'frame through which contemporary political injustices can be observed and examined' (Sugiyama, 2015, p. 176). In other words, the answer to the question of who should have 'a right to have rights' (cf. Arendt, 1958, p. 298; Wolfe, 2012, p. 7), is constantly evolving and Wolfe encourages a critical scrutiny of how lines are drawn and how they could be drawn differently. In relation to the way the distinctions are made between different forms of lives, contemporary biopolitics, according to Wolfe, includes drawing 'not one line ... but many' (Wolfe, 2012, p. 74).

The approach in this book is similar. It acknowledges that one of the central features of neoliberal biopolitics is that it distinguishes and makes divisions between forms of lives, between humans and other animals. Agamben's terms *zoë* and *bios* are important to the methodological approach of this book, as are postcolonial and posthumanist insights into the way these distinctions are made. In relation to postcolonial insights, we shall see that one key to recognising how the notion of population is tied to colonial hierarchies is to deepen the understanding of what Thomas Malthus and (neo)Malthusian logics have meant, and continue to mean, in relation to the development of liberalism more generally (Dean, 2015; Tellmann, 2013) and in relation to environmental discourse more specifically.

In this book the biopolitical questions in relation to water governance revolve around the distinctions made between different forms of lives in liberal biopolitics within a sustainable development regime. This also includes exploring its effects in terms of how water users view themselves, make sense of, and resist, the hierarchies produced in such strategies of governing. Yet, how such mechanisms of inclusion and exclusion function in relation to policies of sustainable development and water management (especially through the notion of the right to [basic] water), is treated as an empirical question and analysed with the help of the above presented biopolitical theories.

Governmentality, biopolitics, and the environment

Up until recently, studies of 'the environment' and environmental governance that use Foucauldian perspectives have been sparse. A seminal publication, *Discourses of the Environment* (Darier, 1999), brought the concepts of biopolitics (Rutherford, 1999a) and governmentality (Luke, 1999) into debates about the study of nature and the construction of 'the environment'. Before this, issues relating to resources, nature, and the environment were only implicit themes in relation to a 'biopolitics of the population', since an explicit focus on 'nature' and the 'environment' is absent from Foucault's analyses (Rutherford, 1999a, p. 45; see also Darier, 1999, p. 6). It has been argued that Foucault did not adequately address the way that nature and the environment became problematized along with the political and the economic problematization of populations (Rutherford, 1999a, p. 45). As we shall see further on in this chapter, a lacuna in Foucault's thought has been identified in that he did not pay attention to the 'Malthus effect' (cf. Dean, 2015) in his genealogy of liberal economic thinking (ibid.; Tellmann, 2013). In any case, early elaborations of 'green governmentality' were explained with reference to Foucault's notion of government as 'the right disposition of things':[7]

> In green governmentality, the disciplinary articulations of sustainability and development centre on establishing and enforcing 'the right disposition of things' between humans and their environment.
>
> (Luke, 1999, p. 146; see also Luke, 1995, p. 69)

In more recent studies, the concept has been slightly modified. Arun Agrawal (2005), who is one of the most well-known green governmentality scholars, for example, places an emphasis on the productive aspects of environmental regulation in terms of the subjectivities that are formed in power relations. He writes in *Environmentality*:

> My use of the term is indebted to Luke for the coinage but it is different in both intent and meaning. I attempt to examine more insistently the shifts in subjectivities that accompany new forms of regulation rather than seeing regulation as an attempt mainly to control or dominate.
>
> (Agrawal, 2005, p. 233)

Furthermore, Timothy W. Luke (1995, 1999) has brought attention to how the development of the technique of biopower can be understood in relation to the 'emergence of "the environment" as a nexus for knowledge formation and as a cluster of power tactics' (1995, p. 66). Luke explains:

> As human beings began to consciously wager their life as a species on the outcomes of these biopolitical strategies and technological systems, it became clear that they also were wagering the lives of other (or all) species as well. While Foucault regards this shift as one of many lacunae in his analysis, it is clear there is much more going on here than he realizes. Once human power/knowledge formations become the foundation of industrial society's economic development, they also become the basis for the physical survival of all terrestrial life forms. Here, ecological analysis emerges as a productive power formation that reinvests human bodies – their means of health, modes of subsistence, and styles of habitation integrating the whole space of existence with bio-historical significance by framing them within their various bio-physical environments filled with various animal and plant bodies.
>
> (Ibid.)

And similarly, Paul Rutherford argues that:

> [I]t is clear from Foucault's discussions of the biopolitical regulation of populations that this assumes not only the disciplining of individuals and populations, but also, necessarily, a concern with the administration of 'all the conditions of life' as represented by the environment.
>
> (1999a, p. 45)

Hence, contrary to much biopolitical scholarship, which focuses on an anthropocentric problematization of biopolitics, there is a great deal to explore here in terms of what biopolitics implies for non-human life, which has been indicated above in relation to contemporary debates on biopolitics.

At present, governmentality and biopolitics theories have been applied to numerous subjects and fields. In addition to the aforementioned water-related research (Bakker, 2012, 2013; Gandy, 2004, 2006; Joyce, 2003; Kooy and Bakker, 2008a, 2008b), the research most relevant to this study includes analyses of sustainable development (Death, 2010; Duffield, 2007; Reid, 2012, 2013; Skoglund and Börjesson, 2014), and of the broader field of 'green governmentality/environmentality/eco-governmentality' (Agrawal, 2005; Birkenholtz, 2009; Darier, 1999, Luke, 1995, 1999; Rutherford, 1999a, 1999b, 2007; Ward, 2013).[8,9]

In this book, a green governmentality perspective is applied to explore how we can understand the relationship between power and the construction of subjectivities in environmental and water governance, as well as how we can comprehend its biopolitics. Such an understanding brings into focus how the relationship between resources and populations can be problematized and how we can view certain notions of scarcity as productive of particular forms of governance.

Problematizing the certainty of governing

One of the focal points of this book is not only about studying the governing techniques per se but also their outcomes in terms of people's lives. Stephanie Rutherford, a green governmentality theorist, has problematized the way scholars, especially in geography, approach the study of 'nature and its rule' from a governmentality perspective (2007, p. 291). While pointing to the productiveness of the governmentality approach in terms of its usefulness for interrogating 'the intersections between nature, power and society' in new ways, Rutherford points to two important critiques of the way that governmentality studies have been conducted in general. First, she notes that the governmentality literature 'needs to be made more complex and attune to difference' (ibid.). In her understanding, there is a tendency in governmentality studies to view 'programmes, policies, practices and techniques of rule ... [as] ... completed projects' (ibid., p. 300). Second (and relatedly), she argues that 'some of the work in this field has ignored the complicated realities of social location, imagining that all bodies are incorporated into systems of rule in the same ways' (ibid., p. 292) Thus, she argues that:

> By paying attention to unintended consequences, acts of resistance, processes of occlusion, and multiple locations in the exercise of green governmentality, I think geographers can provide a more nuanced

> picture of how rule circulates and indeed, is changed by this circulation.
>
> (Ibid., p. 305)

Taking Rutherford seriously means that we need to treat the success, and effects, of governmental rule as an empirical question. This involves making the studies of environmentality 'messier' (ibid., p. 300), and taking into account the specific characteristics of local settings. Such a local example in terms of a specific implementation of a 'hydromentality' is discussed in Chapter 5.

Environmental subjectivities/the greening of the self

Another central theme of green governmentality studies that is important for this book is the recognition of how environmental politics/governance works by calling upon the subjects' ethics and morals (which is elaborated on further in this chapter) and of how environmental politics is productive of certain subjectivities (see Agrawal, 2005; Nightingale, 2011; Nightingale and Ojha, 2013).

Environmental discourses, implicitly and explicitly, bring forward the idea that if people do not 'conduct themselves properly' they will both degrade the environment and bring about their own misery. As is outlined further on in this chapter, this argument has been especially prominent in relation to the poor. However, it can be argued that the environment, environmental 'problems', and resource use are issues that call upon *all* to act, sustainably and responsibly. Luke explains:

> Many individuals who are intent upon turning the world into 'a better place to live' often turn today to 'the environment' in order to make their improvements. Believing that they must do anything and everything to protect 'the environment', they transform this undertaking into a moral crusade.
>
> (Luke, 1995, p. 59)

Those actions can rely 'on a discourse of self-limitation and self-denial', as Sandilands argues (1999, p. 80). In a neoliberal era, people are called upon as consumers, requesting them to consume *properly* (products that are produced sustainably), rather than *less*, in order to become more 'green'.

Such a focus on environmental subjectivities or 'greening' of the self is a central theme in a Foucauldian governmentality analysis.[10] Foucault argued that four aspects (1997, p. 265) are relevant to comprehending the individual's work when turning the self into an ethical subject. These aspects include, first of all, the part of the self or the behaviour that is 'concerned with moral conduct'; second, the ways in which people are invited to 'recognise their moral obligations'; third, the means through which they can change themselves to become ethical subjects; and fourth, the kind of being we aspire to be when behaving 'in a moral way' (ibid., pp. 263–265; see also Hansson and Hellberg, 2015).

Sylvia Bowerbank has elaborated on Foucauldian notions of the subject, subject constitution, and technologies of the self, to explore the work that people do on themselves with the aim of becoming 'green'. In order to understand 'the practices used to define and to change the self in light of ecological principles', she has analysed what she calls 'nature writing' in 'nature journals' and notebooks where green subjectivity is narrativized (1999, p. 164).[11] Bowerbank discusses the productive effects of guilt and truth telling/confession in the narrativization of nature and what people understand as ecological wrongdoing. In Bowerbank's work the act of narrativization is part of a self-transformation to become more environmentally minded (Mikulak, 2013). A guilty conscience, Bowerbank notes (and here she builds on McWhorter, 1992), should not be taken as a guarantee for reformed habits, but could be seen as a technology that can protect the guilty, as self-defence against a call for change (Bowerbank 1999). Relatedly, confession, Bowerbank writes, 'is well entrenched in Western Culture as a disciplinary mode of self-reformation' (1999, p. 169). Foucault explains:

> The confession is a ritual of discourse in which the speaking subject is also the subject at statement; it is also a ritual that unfolds within a power relationship, for one does not confess without the presence (or virtual presence) of a partner who is not simply the interlocutor but the authority who requires the confession, prescribes and appreciates it, and intervenes in order to judge, punish, forgive, console and reconcile; a ritual in which the truth is corroborated by the obstacles and resistances it has had to surmount in order to be formulated; and finally, a ritual in which the expression alone, independently of its external consequences, produces intrinsic modifications in the person who articulates it: it exonerates, redeems, and purifies him, it unburdens him of his wrongs, liberates him, and promises him salvation.
>
> (Foucault, 1998, pp. 61–62)

Hence, according to such a perspective, rituals of truth telling and confession (see also Foucault, 2010, 2011) can be understood as ways for the narrator (in terms of green subjectivities) to handle the ethical puzzles of his or her relationship to nature/the environment. In relation to water use, we will see further on in this book that subjects work on themselves and construct themselves as moral beings, particularly due to water's characteristic as a scarce and vital resource. In the case of eThekwini, we also explore how *responsibilization*, in the sense of making individuals responsible through their individual choices (Rose, 1996, p. 57; see also Hansson, 2015), is made use of in strategies of governing. Such processes of responsibilization are a central neoliberal governing technique. Moreover, we look into how water users construct themselves as (ir)responsible and (un) sustainable in relation to such ways of governing in water services delivery.

(Green) governmentality and the biopolitics of resource governance

This section takes a step back to scrutinize the underlying assumptions and stipulations of environmental/resource governance discourses. It uses the insights of Foucault and more recent theoretical developments of governmentality, biopolitics, and the specific field of green governmentality.

The environment–population nexus is discussed in relation to the notion of scarcity and how scarcity and limits have been understood in connection to different populations. It also outlines the notion of basic needs and the way that regulation in a sustainable development regime envisions particular kinds of subjects. Through this discussion, it addresses liberal aspects of these management approaches as understood from a governmentality perspective.

Taken together, what we learn from the section below is that in order to critically scrutinize water management from a biopolitical/(green) governmentality perspective, we need to look into the ways in which populations are constructed and governed, how scarcity works as an idea in the government of these different populations, and how the subjects of government are viewed, both in terms of the assumptions of what they are and in relation to what they are envisioned as becoming.

The population–resource nexus

As already indicated in the presentation of the notion of biopolitics, the 'problem of the population' is fundamental to modern governmentality. The population as a focus of government is connected to the development of statistics in the eighteenth century. Then statistical measures, for the first time, made it possible to gain knowledge of the population and 'its own regularities, its own rate of deaths and diseases, its cycles of scarcity, etc.' (Foucault, 1991, p. 99). The problem of the population is also absolutely central in discussions on the environment and resource management. Rutherford, a green governmentality theorist who builds on Foucault, explains that when *population* became the central economic and political problem, the focus on government also became a question of the *relationship between population and resource availability*. He explains:

> in this process, not only does the idea of a measurable and manageable population come into existence, but also does the notion of the environment as the sum of the physical resources on which the population depends.
>
> (1999a, p. 39, italics in original)

During large parts of the twentieth century, political discourse about the environment was embedded in a view of nature as a source of 'goods' for human society. Nature, from such a perspective, is reduced to a system of 'biophysical systems that can be dismantled, redesigned, and assembled anew to produce "resources" efficiently and in adequate amounts when and where needed in the modern marketplace' (Luke, 1995, pp. 70–71). According to Luke (ibid., p. 70), 'resource managerialism can be read as the eco-knowledge of modern governmentality'.

In relation to water management, the next chapter looks at how resource management was especially prominent in the so-called hydraulic mission paradigm, but can be argued to still be alive and kicking in present-day approaches such as Integrated Water Resources Management (IWRM), according to critical scholars such as Linton (2010). Through the birth of hydrology and the construction of the hydrological cycle, water has been represented as a 'natural fact' (Linton, 2010, p. 184) that can be made readily available for human use.

The relationships between population and resources are, however, not as straightforward as they may seem as they take shape in (neo)liberal governmentality. Below is an account of how we can make sense of the relationship between (different) populations, resources, and scarcity in a liberal economy as seen from a Foucauldian perspective. What is important here, however, is to pay particular attention (more than did Foucault himself) to the ways in which Malthus' formulation of the 'principle of the population' have influenced these rationalities (Dean, 1991, 2015; Tellmann, 2013). This provides an idea of how certain ideas of scarcity, and how it is to be avoided, can function so as to govern populations and individuals. Moreover, this discussion introduces how *moralism* features in environmentality in terms of how it requests people to 'conduct themselves properly' (ibid., p. 43) in order to avoid 'misery and vice' (Malthus, 1798). As we will see further on in this book, this is a theme that becomes central when exploring both how governments target water users and how water users productively engage with governing logics.

Liberalism, scarcity, and population

In *Security, Territory, Population,* Foucault describes in detail the shift from a mercantile economy to a system that allows a 'free circulation of the grain' in seventeenth- and eighteenth-century France. It is simultaneously a discussion of the notion of scarcity and a description of the shift from a juridical/disciplinary society to a system that directs its attention to population regulation, which Foucault calls a:

> technique of apparatuses of security … [and] states: You could read the principle of the free circulation of grain … as an episode in the mutation of technologies of power and an episode in the instalment of

> this technique of apparatuses of security that seems to me to be one of the typical features of modern societies.
>
> (2007, p. 34)

The earlier (juridical and disciplinary) system is described by Foucault as an 'anti-scarcity' system' (ibid., p. 32). In this system, scarcity was something that needed to be avoided because of the risk of revolt in urban areas when prices rose, which, in turn, could lead to a crisis or catastrophe for the government (ibid., p. 30). To prevent scarcity, the system controlled prices and the right to store grain, and it put limits on exports with the aim of attaining the lowest possible price so that the town could be fed at the lowest cost (ibid., p. 32). Thus, the state controlled, constrained, and supervised the production and trade of grain (ibid.).

However, economic thinkers of mid-eighteenth century France (i.e. the physiocrats) gradually began to understand 'the free circulation of the grain' as the better system for avoiding scarcity (ibid., p. 33) since market forces should compensate for fluctuations in production. Up until then, scarcity had been:

> both an individual and collective phenomenon, and in the same way that people suffered hunger, so entire populations and the nation suffered hunger, and it was precisely this kind of immediate solidarity, the massiveness of the event, that constituted its character as a scourge.
>
> (Ibid., p. 41)

In the new system, scarcity becomes something of a 'chimera' (ibid.). In a general sense, *on the level of the population*, scarcity disappears. Yet this does not mean that some will not experience difficulties in buying wheat and 'it may well be that some people die of hunger after all' (ibid., p. 42). Foucault explains, however, that this shift means that the politico-economic rationalities are merely interested in the individual insofar as she or he is an instrument to achieve goals at the level of the population:

> we will have an absolutely fundamental caesura between a level that is pertinent for the government's economic-political action, and this is the level of the population, and a different level, which will be that of the series, the multiplicity of individuals, who will not be pertinent, or rather who will only be pertinent to the extent that, properly managed, maintained, and encouraged, it will make possible what one wants to obtain at the level that is pertinent.... The final objective is the population.
>
> (Ibid.)

The population emerges as a (or *the*) political subject, a new collective subject, which is different to the (individual) political subject in the juridical/disciplinary system. In the disciplinary system, the norm was the starting point and individual behaviour was to be corrected to conform to this norm. The disciplinary system was thus concerned with *normation* rather than *normalization.* In this new system, government is not directed towards the individual in terms of what the individual is allowed to do and what is prohibited according to the law. In a system that Foucault describes as the 'apparatuses of security':

> instead, we have a plotting of the normal and the abnormal, of different curves of normality, and the operation of normalization consists in establishing an interplay between these different distributions of normality and [in] acting to bring the most unfavourable in line with the more favourable.
>
> (Ibid., p. 63)

Thus, while discipline attends to detail, regulation of the population/security 'lets things happen' (ibid., p. 45). According to Foucault, it is in relation to 'this option of the circulation' that we should understand the word freedom in liberalism (ibid., p. 49). To Foucault, this is '[t]he game of liberalism' – not interfering, allowing free movement, letting things follow their course (ibid., p. 48). Ultimately, for Foucault, this freedom is a central dimension of the deployment of 'apparatuses of security' (ibid., p. 49).

Furthermore, the system is based on the premise that people 'conduct themselves properly' (ibid., p. 43), which means that they accept the rules of the game and act accordingly. In the example of the grain, it is required that some sell their grain at the right moment, some endure scarcity, and others export when prices rise. Such individuals are well-functioning elements of the population – they 'act as members of the population' (ibid.). Those who resist doing so, those who act contrary to the system (and Foucault refers here to a text by the physiocrat Abeille) can be understood as placing themselves *outside* the population. They are simply 'the people' (ibid.). Those who do not conduct themselves properly – and in this case suffer from hunger as a consequence – can be understood as people who 'bring about their own misery'. Here, a kind of moralism features in the way that people should take responsibility for, and act in accordance with, what they are asked to do as actors within the larger system.

Chapter 5 shows how such logics resonate with the hydromentality of eThekwini in relation to water users who connect illegally to the water grid. Such individuals can, from the municipality's point of view, be regarded as placing themselves outside the population through acts that are contrary to the system. We will also see how similar ideas appear in the water users' stories. In some narratives, such logics echo in the way that water users speak of others who fail to 'conduct themselves properly' and who do not take enough care of how much water they are using so that they risk having their water access cut off (those who rely on limited access), or end up with large debts (those who buy water on credit).

This kind of moralism corresponds well to the legacy of Malthus and his thesis that the poor bring about their own suffering. Dean (1991, 2015) and Tellmann (2013) note, however, that a proper engagement with Malthus is lacking in Foucault's discussion on biopolitics. Therefore, Foucault's writings about the relationship between populations, resources, and scarcity in liberalism, which can be used for analysing sustainable development and global water governance, need to be complemented with a focus on Malthus, who, as is convincingly argued by Tellmann, modified the links between the biopolitics of the population and liberal economy. Since Malthus' population principle involved a shift in liberal population discourse, the omission of Malthus in Foucault's writing and lecturing is 'curious' (Tellmann, 2013, p. 136) and it 'makes liberalism look different – even more different than Foucault made it out to be' (ibid., p. 152).

The 'Malthus effect' and different populations

In environmental discourse, it is widely recognised that Malthus' *An Essay on the Principle of Population* (1798) has been one of the most influential publications when it comes to setting the stage for the discussion on population and its relationship to resources. In his essay Malthus stipulated that population growth increases faster than its means of subsistence but is 'kept equal to the means of subsistence, by misery and vice' (Malthus, 1798). Malthus thus introduced the idea of 'a constant and pressing disequilibrium between population and subsistence' (Dean, 1991, p. 120), of a looming catastrophe through the threat of scarcity (Dean, 1991; Tellmann, 2013, building on Halévy, 1995). Hence, as Tellmann argues, Malthus' publication 'raised the spectre of overpopulation and made the notion of scarcity central for liberal economic discourse in ways that it had not been before' (2013, p. 136).

Green governmentality theorists have previously outlined how Malthus' legacy has created the notion that there is a 'possibility that there are, will or could be too many people than is good for us' (Sandilands, 1999, p. 82). In environmental discourse there is a general belief that there is a negative relation between population growth and environmental degradation. Sandilands explains that 'for many environmentalists, the link between population growth and environmental degradation seems so obvious as to be a truism' (ibid., p. 84).

The question of population growth in environmental discourse can be analysed productively from a biopolitical perspective in terms of how this population discourse distinguishes between different forms of lives. Hence, there are *certain* populations that are viewed as the problem in relation to their effects on the environment and resource use; the Malthusian population principle produced a fear that the poor and 'underdeveloped', that is, 'savage life' as Tellmann writes, would 'eat up the future for all' (2013, p. 151).

This fear is produced through a colonial imaginary that was central to Malthus' framing of the population question; an imaginary that involved 'a colonial hierarchy that differentiates between dangerous '"savage" and economic "civilized" life' (ibid., p. 135). This hierarchization, in turn, is based on the different relationships that these populations have to economic futurity (ibid., p. 137), which ultimately explains their success, or lack of it, in terms of historical and material progression (ibid., p. 143).

Malthus' principle of population explains why certain populations do not reach the stage of civilization; it is due to the immediacy with which they react to a situation of abundant resources. Just like animals, the 'uncivilized states of man' react to an abundance of resources with 'a large proportion of the procreative power' (Malthus, cited in Tellmann, 2013, p. 144). In turn, this rapid large growth in population is checked by 'violent causes', (ibid.) leading to population decline.

The central argument here is that just because 'savage life', occupied as it is with living in the present without a sense of either history or future (Tellmann, 2013, building on Spurr, 1994), is not able to turn the present abundance into a sustainable future, it becomes locked into 'an endless circle of abundance and destruction' (Tellmann, 2013, p. 144). In Tellman's words the figure of the population, as it appears in Malthus' principle:

> is constitutively determined by an inner 'ladder of hierarchizations' that differentiates between more animal, more savage and more catastrophic forms of life on the one hand and more human, more civilized and more economic forms of life on the other.
>
> (Ibid., p. 151)

Important for the argument of this book is that in this context, scarcity is not (only) about a situation in which needs exceed resources. Neither is scarcity, as it takes shape here, about the 'limits of the earth' (ibid., p. 145). Rather, scarcity becomes something of a 'regulatory device' (ibid., p. 137) that is meant to unlock savage life from its occupation of living only in the present. This involves a reorientation towards acting '"as if" scarcity prevails in any given moment' (ibid., p. 146) and through which the future will be taken into account. Hence, scarcity has an 'uncertain status.... It wavers between what is given and what is not yet; between what is and what should be apprehended and averted' (ibid.).

What is especially important to note here, in relation to the theme of this book, are the hierarchizations of populations, the 'biopolitical caesuras' (ibid, p. 142), in relation to the way that populations are thought and governed. Equally important is how the notion of scarcity becomes something of a technology of governing. Both these aspects are still central to governing rationales in a sustainable development regime, as is shown below. Hence, it is important to take into account Malthus' legacy both for tracing the genealogy of liberalism (Tellmann, 2013, p. 152) and for understanding it in its contemporary shape. In concurrence with this, Dean argues that '[t]he Malthusian concern with the population as the unstable equilibrium between its growth and environmental resources has thus demonstrated a remarkable persistence over more than two centuries' (2015, p. 34).

Malthus' legacy has, however, in the course of time taken on 'different political, epistemic and normative guises' (ibid.). One example of this is the way that the idea of 'overpopulation' came to be a condition of the formation of the environmental movement in the 1960s (ibid., p. 30). Through figures such as Paul Erhlich and his publication *The Population Bomb*, overpopulation became portrayed as the root of social, economic, and environmental problems (ibid.).

However, a shift in relation to the original Malthusian ideas can be detected in such neo-Malthusian approaches. While for Malthus it was the consequences of reproduction for *people* that were the focus, for the neo-Malthusians it is the effects on the *environment* (and thereby indirectly on people) when an increasing number of people overexploit resources (Gausset *et al.*, 2005). In this way it is not only the number of people that is formulated as a problem but also affluent populations and their use of resources and creation of waste (Dean, 2015, p. 31). Furthermore, in contrast to Malthus, who was optimistic about the possibility that production and thereby population could continue to increase (ibid.), making absolute limits of the earth 'irrelevant' to the problem of population (Tellmann, 2013, p. 145), neo-Malthusian approaches also produced the understanding that indefinite growth in a finite world is unsustainable.

This was also the message in the Club of Rome's widely debated publication *The Limits to Growth* (Meadows *et al.*, 1972, p. 23), in which the negative relationship between population growth and resource use was utterly clear. It stipulated that:

> if the present growth trends in world population, industrialization, pollution, food production, and resource depletion continue unchanged, the limits to growth on this planet will be reached sometime within the next one-hundred years.
>
> (Ibid.)

The publication spurred a contentious debate over the relationship between the resource base of the earth and human demands, and over the very central themes of growth, limits, and scarcity (see, for example, Boserup, 1973; Daly, 1977; Hornborg, 2003; Lomborg, 2001). Critics argued that the limits-to-growth thesis simplified complex relationships between livelihood strategies and resources available and that technological innovation was downplayed (Gausset *et al.*, 2005; Pirages, 2005). There was not only critique in terms of the methodology of the limits-to-growth thesis, but also concerning its doomsday rhetoric.[12,13] Neoliberal thinkers, such as Friedrich Hayek, and in the United States, particularly through the Reagan administration, portrayed *The Limits to Growth* report, and neo-Malthusian environmentalists in general, as an anti-progress movement (Dean, 2015, p. 33, building on Greene, 1999).

In a more general sense, environmental discourses that drew on the limits-to-growth thesis were understood as a critique of modernization and conventional economic strategies of development. Somewhat simplified, environmental discourse in the 1970s and 1980s perceived the relationship between a concern for the environment and development in negative terms. However, as environmental/resource concerns gradually entered mainstream discussions and became manifest in the concept of sustainable development, the relationship between the environment and development was reformulated.

Compared to modernization critiques during the 1960s and 1970s, in publications such as *Silent Spring* (Carson, 1962) and The Club of Rome's report *The Limits to Growth* (Meadows *et al.*, 1972), the view of sustainable development shifted. This shift meant that environment and development were no longer seen as opposing categories (environment *or* development). Instead, a perception in which a care for the environment and for development is possible simultaneously (environment *and* development) became dominant, and even the idea that the two categories are mutually reinforcing (environment *for* development) (UNEP, 2007, p. 10).

Merging environmental and developmental concerns like this affects the way that the discourse on sustainable development frames issues such as limits, scarcity, growth, and poverty. As we shall see, some of Malthus' original stipulations resurface through the notion of sustainable development, particularly in relation to poverty and the way that different populations are understood.

Sustainable development discourse, limits, and growth

Sustainable development is known as 'development that meets the needs of the present, without compromising the ability of future generations to meet their own needs' (WCED, 1987, chapter 2, paragraph 1). In the Brundtland report, the approach to limits to resources and waste absorption is formulated as follows:

> The concept of sustainable development does imply limits – not absolute limits but *limitations* imposed by the present state of technology and social organization on environmental resources and by the ability of the biosphere to absorb the effects of human activities. But technology and social organization can be both managed and improved to make way for a new era of economic growth.
>
> (WCED, 1987, part 1, paragraph 27, my italics)

Growth, or *sustainable growth*, was thus defined as a prerequisite for dealing with environmental issues. Thereby sustainable development successfully unites a care for the environment with a concern for 'sound' economic policy. The conviction that economic growth and technology are solutions to environmental challenges is still at the heart of mainstream environmental discourse, which can be illustrated by the closing words of the narrator in the video clip 'Welcome to the Anthropocene' that opened the Rio+20 conference in 2012:

> We have entered the Anthropocene, a new geological epoch dominated by humanity. This relentless pressure on our planet risks unprecedented destabilization. *Our creativity, energy, and industry offer hope.* We have shaped our past. We are shaping our present. We can shape our future. You and I are part of the story. We are the first generation to realize this new responsibility. *As the population grows to 9 billion* we must find a safe operating space for humanity, for the sake of future generations.
>
> (Planet under pressure, 2012, my italics)

The quotation reiterates the belief in humanity's capacity to overcome these challenges through innovation and technology. Such an approach to environmental issues has been labelled as Ecomodernization and has been criticized for merely promoting business as usual. Some critics understand sustainable development as nothing more than mainstream capitalist economic strategies taking on a new language of the environment, a kind of 'Sustainababble' (Hornborg, 2003), which does not take seriously the limits of the earth's resource base and its capacity to absorb waste. In the report that was the outcome of the Rio+20 meeting '[t]here is no mention of planetary boundaries, limits or carrying capacity of the world' (Clémençon, 2012, p. 317). This, Raymond Clémençon argues, has been a disappointment for environmental groups, since science has emphasized the need to embrace upper limits on greenhouse gases to avoid escalating and irreversible climate change, and thus reflects 'a disturbing science-policy disconnect' (ibid.). The belief in growth and technological innovation so central to sustainable development is also related to the view of poverty that was initially portrayed in the sustainable development discourse, since poverty was framed as both a cause and an effect of environmental degradation.

Sustainable development discourse and different populations

In the early documents and debates around sustainable development, the relationship between poverty and environmental interactions was discussed in terms of a 'downward spiral'. According to such views, the poor were assumed to be placing an increasing pressure on resources. In turn, the environmental degradation following these pressures was assumed to be leading to increasing poverty (Scherr, 2000, p. 481). In these descriptions, the poor are portrayed as lacking the capacity to act responsibly and sustainably (due to their lack of alternatives), a problem that is projected to accelerate with population growth in developing countries.[14]

In the Brundtland report, poverty was explicitly defined as both a cause and an effect of environmental degradation. The report states that:

> Poverty is not only an evil in itself, but sustainable development requires meeting the basic needs of all and extending to all the opportunity to fulfill their aspirations for a better life. A world in which poverty is endemic will always be prone to ecological and other catastrophes.
>
> (WCED, 1987, From One Earth to One World, paragraph 27)

> Those who are poor and hungry will often destroy the immediate environment in order to survive: They will cut down forests; their livestock will overgraze grasslands; they will overuse marginal land; and in growing numbers they will crowd into congested cities. The cumulative effect of these changes is so far reaching as to make poverty itself a major global scourge.
>
> (Ibid., chapter 1, paragraph 8)

In the report, a negative relationship between poverty and sustainable development is thus assumed, while a positive relationship between economic growth and environmental care is established. As we recall, these arguments relate to the way in which Malthus outlined the differences between savage and civilized life and their relation to futurity. In economic terms, this is formulated in the stipulation that the poor have 'a higher rate of time preference' (Moseley, 1991, p. 317) compared to wealthier individuals.

In accordance with Malthus' thesis, this means that the poor, or so it is assumed, are preoccupied with surviving the present rather than saving for the future. However, as noted above, in environmental discourse the concern is less about the effects on reproduction than about how these mechanisms lead to an overexploitation of the environment. In line with this thesis, the opposite is true for wealthy individuals, who can invest in the future and environmental conservation (ibid.). This view produced policy responses such as population growth control, control of resource access and use by the poor, and environmental education (Scherr, 2000, p. 481).

Research has, however, long shown that poor individuals and households behave in quite a contrary manner to the thesis of higher time preference. Rather than just thinking of how to survive the day, they undertake measures – sometimes such extreme measures as depriving the family of needed calories (Moseley, 1991) – in order to guard against future shortages (see also Scherr, 2000; Templeton and Scherr, 1999).

Despite understandings of the relationships between poverty and the environment having been made more complex within research and the fact that there has been a shift in terms of how the relationship between poverty and environment is portrayed, the stipulation that poverty is a main generator of environmental degradation is still strong in environmental discourse. Consequently, the population–environment nexus is an important part of 'environmentality' (Sandilands, 1999, p. 84) and, as we shall see further on in this book, informs the ways that resource management targets (different) populations in a sustainable development regime. Furthermore, poverty as a generator of environmental degradation and wasteful resource use is also manifest in the ways that the water users themselves speak of these issues, as is shown in Chapter 5. Nevertheless, the narratives also testify to how a poor background produces a frugal use of (water) resources.

Sustainable development and basic human needs

Another important aspect of sustainable development, especially from a biopolitical perspective, is its focus on (basic) needs. The goal of satisfying (basic) needs is implied in the definition of the concept and might seem just and straightforward. The Brundtland report states, for example, that '[t]he satisfaction of human needs and aspirations is so obviously an objective of productive activity that it may appear redundant to assert its central role in the concept of sustainable development' (WCED, 1987, chapter 2, paragraph 42). Employment, food, energy, housing, water supply, sanitation, and health care are mentioned as basic human needs (Holden and Linnerud, 2007; WCED, 1987), and the report asserts that 'overriding priority should be given' to 'the essential needs of the world's poor' (WCED, 1987, Conclusion).

However, as Holden and Linnerud (2007) note, the Brundtland report also recognises people's desires for more than merely meeting their essential needs. The report states: 'Sustainable development requires meeting the basic needs of all and extending to all the opportunity to satisfy their aspirations for a better life' (WCED, 1987, chapter 2, paragraph 27). In order to be in accordance with the overall goal of the concept, such aspirations need to be ecologically sustainable, however, implying that 'not every aspiration for a better life is compatible with the goal of sustainable development' (Holden and Linnerud, 2007, p. 177).

The presentation of the two key concepts in the Brundtland report, 'needs' and 'limitations', of which the latter are 'imposed by the state of technology and social organization on the environment's ability to meet present and future needs' (WCED, 1987, chapter 2, paragraph 1), is telling in relation to the anthropocentrism of sustainable development. Thereby, environmental sustainability becomes relevant to the extent that it has an effect on the possibility for present and future generations to meet their needs and their aspiration for a better life. The report concludes:

> In essence, sustainable development is a process of change in which the exploitation of resources, the direction of investments, the orientation of technological development, and institutional change are all in harmony and enhance both current and future potential to meet human needs and aspirations.
>
> (Ibid., paragraph 15)

The distinction between meeting basic needs and the aspiration for a good life is central to biopolitics. According to what has been previously outlined in this chapter, it can be understood as the difference between *bios* and *zoë*; between life as mere survival and qualified life; between simple living and proper life (cf. Agamben, 1998; Esposito, 2008a). The question is: who can strive for a better life in the biopolitical frame that sustainable development entails? Mark Duffield has described the entry of sustainable development to mainstream development thinking and practice as a deepening and confirmation of the 'biopolitical division in global population between insured and non-insured peoples' (2007, p. 68). He argues that:

> Rather than reducing the life-chance divide between the developed and underdeveloped world, sustainable development is better understood as a means of containing the latter. Poor and non-insured communities are expected to live within the limits of their own powers of self-reliance
>
> (Ibid.)

and be content with satisfying their basic needs (ibid., p. 70).

In Duffield's work, sustainable development functions as an extension of neoliberal principles of economy, as it involves a break with state-led modernization strategies that shifts the burden of security from states to people (ibid., p. 69). According to this perspective, sustainable development promotes a neoliberal model of society in which everyone is required to 'prove themselves by bettering their individual and collective self-reliance' (ibid., p. 69; see also Hellberg and Knutsson, 2016). In Chapters 3 and 5, the discussion on basic needs is expanded in terms of how we can understand *notions of different populations* and *the right to basic water* from a biopolitical perspective. Such a policy of granting people a minimum level of security (cf. Foucault, 2008, p. 207) can be analysed in terms of governing strategies that allow the market to function, rather than being counterpoints to such principles. In contemporary sustainable development policy discourse, this element of self-reliance has been further strengthened through the promotion of *resilience*, to which we will now turn before concluding the chapter.

Sustainable development and resilience

The establishment of resilience as a central concept involves a shift in environmental governance discourse. While previous discourses around environment and sustainable development relied on the assumption that 'everything it stipulates can be known' (Luke, 1999, p. 138), a focus on resilience emphasizes uncertainty.

This shift entails a change in terms of the ontologies of both the environment and human subjects. Within a discourse of resilience the environment is understood as uncertain and risky and humans must become adaptive and self-reliant in relation to these uncertainties and risks (Duffield, 2007; Reid, 2012, 2013). This new commitment to resilience in global governance institutions is, for example, manifest in a joint report from the World Resources Institute and the United Nations Development and Environmental Programmes (WRI/UNDP/UNEP), *Roots of Resilience* (World Resources Institute, 2008). In this report the sustainability concept appears in relation to the three dimensions associated with the concept of sustainable development: economic, social, and ecological. The report states:

> Resilience is the capacity to adapt and to thrive in the face of challenge. This report contends that when the poor successfully (and sustainably) scale up ecosystem-based enterprises, their resilience can increase in three dimensions: They can become more economically resilient – better able to face economic risks. They – and their communities – can become more socially resilient – better able to work together for mutual benefit. And the ecosystems they live in can become more biologically resilient – more productive and stable.
>
> (World Resources Institute, 2008, ix)

In accordance with this, a resilient approach 'does not require a precise capacity to predict the future, but only a qualitative capacity to devise systems that can absorb and accommodate future events in whatever unexpected form they may take' (Holling, 1973, cited in Walker and Cooper, 2011, p. 146).

There has been critique of the neoliberal biopolitics of the, so-called, sustainable development–resilience nexus. This critique not only addresses how the discourse of resilience legitimates neoliberal forms of governance, but also how it invokes a certain kind of subject (Evans and Reid, 2014; Reid, 2012, 2013). The incorporation of resilience as a key concept in environmental discourse is argued to have the effect of pushing even more strongly for neoliberal governance reforms. This is because of how the market, and its production of economic diversity, becomes a 'core constituent' of the resilience that is to be created. In this process, resilience becomes a property not only of the biosphere but also of humans. Simultaneously, notes Julian Reid, building on Folke *et al.* (2002), the biosphere is reformulated from being an extra-economy domain, which is supposed to be protected from the economic practice of humans, to becoming 'an economy of "services" which "humanity receives"' (Folke *et al.*, 2002, p. 437). The subject that is envisioned, in turn, is a subject that is equipped with agency and capable of self-organization at the same time as being adaptive and ready for change. Reid writes:

> The resilient subject is a subject which must permanently struggle to accommodate itself to the world. Not a political subject which can conceive of changing the world, its structure and conditions of possibility ... [b]ut a subject which accepts the disastrousness of the world it lives in as a condition for partaking of that world and which accepts the necessity of the injunction to change itself in correspondence with the threats and dangers now presupposed as endemic.
>
> (Reid, 2012, p. 74)

Even though resilience is a property demanded of all populations, the poor are the ones who are targeted as in need of a resilient approach (ibid.). This is because they are often the ones who are most directly dependent on ecosystem services – they are 'the wealth of the poor' (UNEP, 2004, p. 10; World Resources Institute, 2005). This focus is also related to ideas on the relationship between the poor and ecological ignorance (also see, for example, Reid, 2012) and the particular form of moralism towards the poor that has been a central theme in environment discourse ever since Malthus, as discussed throughout this chapter.

The following chapter explores the meaning of resilience as it takes shape within the water debate and argues that even though we should be wary of this concept for the reasons outlined by Reid (2012, 2013) and Evans and Reid (2014), the efforts made in its name do not necessarily always mean that the poor will be left worse off. The issue, however, has a particular biopolitical logic to it, which is explored further in the next chapter.

Conclusion

This chapter has presented the theoretical foundations for this book. It has outlined what a governmentality and biopolitical perspective entails for the study of sustainable development and resource governance. It has, furthermore, provided an approach to how we can conceptualize both (neo)liberalism and the workings of power within a (green) governmentality framework. Such an approach views neoliberalism as a practical art of government rather than a political philosophy and emphasizes both the productivity of power relations and the uncertainty of governing.

The framework provides theoretical insights into how neoliberal techniques of governing use the agency of free subjects in order to govern. Processes of responsibilization, together with the moralism argued to be a central component of discourses around sustainable development and care for the environment, are related to how subjects work on themselves through techniques of the self.

Particular focus has been placed on the concept of biopolitics and how it has been developed by philosophers and political theorists such as Agamben, Esposito, and Wolfe. What is central is the emphasis placed by these theorists on the distinctions made between different forms of life, not only between different forms of human life but also between humans and (other) animals (cf. Wolfe, 2012).

The section focusing on the biopolitics of resource governance provided a perspective on the population–resource nexus as seen through a governmentality lens. Through a particular focus on the 'Malthus effect', the chapter outlined how the notion of scarcity – rather than simply depicting a situation in which demand is greater than resources availability – instead functions as a regulatory device, a governing technique in (neo)liberal strategies of governing. Central to this is how such an understanding, as produced by Malthus, is related to a colonial hierarchy between different forms of life, a difference between savage and civilized life which is still alive today – although in a different terminology – in environmental and resource discourses. Using these insights into how liberal government has approached questions of the relationship between populations and resources as a backdrop, in its final sections the chapter explored the meanings of concepts such as limits, growth, and scarcity in sustainable development discourse. It also explored how different populations, that is, poor and rich, are understood in relation to a care for the environment.

The chapter also showed how the focus on basic needs, at the core of the concept of sustainable development, raises central biopolitical problems in terms of the difference between mere life and the good life, and who can aspire to the latter. The focus on basic needs also displays the anthropocentrism of the concept, since ecological sustainability is simply described as a means of meeting basic human needs and aspiring to better lifestyles. Hence, nature and forms of life other than human, emerge as a resource base for economic development and human well-being.

The chapter ended with a discussion on the incorporation of resilience in sustainable development discourse, which has, as argued, strengthened the neoliberal aspects of sustainable development as it places the responsibility of coping with a dangerous and uncertain world on the capacities of (poor) individuals and, in this process, produces a 'politically debased subject' (Reid, 2012).

Notes

1 In contrast, Foucault described a situation when all such possibilities are closed off – as they are, for example, for a slave in chains – as a 'relationship of violence' (Foucault, 2000, p. 340).
2 Foucault also elaborated on another technique of power: pastoral power, associated with early and medieval Christianity. Since the framework developed in this book is concerned with power techniques within the frame of modern governmentality, pastoral power is not a central concept, while recognising that, for a better understanding of the genealogy of Foucault's conceptualization of power, it is fruitful to examine how such a form of power is connected to the exercise of power in modern societies.
3 Following Foucault's description, disciplinary power emerged in the seventeenth and eighteenth centuries to enable the usage of workers' bodies for industrial production while minimizing disobedience (Foucault, 1977, p. 138).
4 To use the term sovereign power rather than sovereignty (understood as 'sovereign statehood') opens up a discussion on sovereign power at a global or international level (Edkins and Pin-Fat, 2004).
5 For contemporary debates on biopolitics and the relationship between sovereign power and biopower, see, for example, Agamben, 1998; Edkins, *et al.*, 2004; Edkins and Pin-Fat, 2005; Neal, 2004; Zevnik, 2009.
6 Judith Butler does not herself place her work under the headline of biopolitics.
7 Foucault builds on the work of de la Perrière who in his definition of government, in contrast to other thinkers of his time, did not refer to 'territory' (Foucault, 1991, p. 93). The full sentence in de la Perrière's words reads: 'government is the right disposition of things, arranged so as to lead to a convenient end' (de la Perrière, 1567, p. 22, referenced in Dupont and Pearce, 2001).
8 This text uses the concepts of 'environmentality' and 'green governmentality' interchangeably.
9 Birkenholtz and Ward explicitly address water governance/management.
10 Joanna Macy has used the phrase 'the greening of the self' in a book which describes to the reader that a new green subject (that is truly affiliated with the earth) is awakening and replacing modern subjectivity (Bowerbank, 1999, p. 165). As Bowerbank notes, in Macy's writings green subjectivity is distinct from a 'modern self'. The green self is in comparison 'true' and 'natural' and in the greening process, 'you become more yourself' (Macy, 1991, p. 189, cited in Bowerbank, 1999, p. 166).
11 Ecology is here understood in a broad sense. Bowerbank builds on Williams (1976) and argues that ecology has moved beyond its initial scientific meaning and includes a concern for human relations with the physical world and the implications for social and economic policy (Bowerbank, 1999, p. 167).
12 However, *The Limits to Growth* acknowledged that the chance existed to 'alter' the trends, and 'to establish a condition of ecological and economic stability that is sustainable far into the future' (Meadows *et al.*, 1972, p. 24). Thus, its prognosis was perhaps not as deterministic as some have argued.
13 For a discussion on doomsday rhetoric in relation to the environment, see also Swyngedouw (2010).
14 The Brundtland report, paragraph 56, reads: 'Pressure on resources increases when people lack alternatives. Development policies must widen people's options for earning a sustainable livelihood, particularly for resource-poor households and in areas under ecological stress' (WCED, 1987, chapter 2, paragraph 56).

References

Agamben, G. (1998) *Homo Sacer: Sovereign Power and Bare Life*. Stanford: Stanford University Press.

Agrawal, A. (2005) *Environmentality: Technologies of Government and the Making of Subjects*. Durham: Duke University Press.

Alt, S. (2016) 'Beyond the Biopolitics of Development: Being, Politics and Worlds', PhD thesis, University of Lapland.

Arendt, H. (1958) *The Origins of Totalitarianism* (2nd edn revised). New York: Meridian.

Bakker, K. (2012) 'Water: Political, Biopolitical, Material', *Social Studies of Science*, 42(4), 616–623.

Bakker, K. (2013) 'Constructing "Public" Water: The World Bank, Urban Water Supply, and the Biopolitics of Development', *Environment and Planning D: Society and Space*, 31(2), 280–300.

Birkenholtz, T. (2009) 'Groundwater Governmentality: hegemony and technologies of resistance in Rajasthan's (India) groundwater governance', *The Geographical Journal*, 175(3), 208–220.

Boserup, E. (1973) *Jordbruksutveckling och befolkningstillväxt [Agricultural development and population growth]*. Lund: Gleerup.

Bowerbank, S. (1999) 'Nature Writing as Self-Technology', in Darier, E. (ed.) *Discourses of the Environment*. Malden, MA: Blackwell Publishers.

Butler, J. (1997) *The Psychic Life of Power: Theories in Subjection*. Stanford: Stanford University Press.

Butler, J. (2006) *Precarious Life: The Powers of Mourning and Violence*. London: Verso.

Butler, J. (2009) *Frames of War: When Is Life Grievable?* London: Verso.

Carson, R. (1962) *Silent Spring*. Boston: Houghton Mifflin.

Clémençon, R. (2012) 'Welcome to the Anthropocene: Rio+20 and the Meaning of Sustainable Development', *The Journal of Environment and Development*, 21(3), 311–338.

Cruikshank, B. (1999) *The Will to Empower: Democratic Citizens and Other subjects*. Ithaca: Cornell University Press.

Daly, H. (1977) *Steady-state economics: The Economics of Biophysical Equilibrium and Moral Growth*. San Francisco: W. H. Freeman.

Darier, E. (ed.) (1999) *Discourses of the Environment*. Malden, MA: Blackwell Publishers.

Dean, M. (1991) *The Constitution of Poverty: Toward a genealogy of liberal governance*. London: Routledge.

Dean, M. (1999) *Governmentality: Power and Rule in Modern Society*. Los Angeles: Sage.

Dean, M. (2015) 'The Malthus Effect: population and the liberal government of life', *Economy and Society*, 44(1), 18–39.

Death, C. (2010) *Governing Sustainable Development: Partnerships, Protests and Power at the World Summit*. London: Routledge.

De la Perrière, G. (1567) *Le Miroir Politique*. Paris.

Duffield, M. (2007) *Development, Security and Unending War: Governing the World of Peoples*. Cambridge: Polity.

Dupont, D. and Pearce, F. (2001) 'Foucault contra Foucault: Rereading the "governmentality" papers', *Theoretical Criminology*, 5(2), 123–158.

Edkins, J. and Pin-Fat, V. (2004) 'Introduction: Life, Power, Resistance' in Edkins, J., Pin-Fat, V., and Shapiro, M. J. (eds.) *Sovereign Lives, Power in Global Politics*. New York: Routledge.

Edkins, J. and Pin-Fat, V. (2005) 'Through the Wire: Relations of Power and Relations of Violence', *Millennium – Journal of International Studies*, 34(1).

Edkins, J., Pin-Fat, V., and Shapiro, M. J. (eds.) (2004) *Sovereign Lives, Power in Global Politics*. New York: Routledge.

Esposito, R. (2008a) 'Immunization and Violence' 1–16. www.biopolitica.unsw.edu.au/sites/all/files/publication_related_files/esposito_immunization_violence.pdf [accessed 18 September 2017].

Esposito, R. (2008b) *Bíos: Biopolitics and Philosophy*. Translated by Timothy Campbell. Minneapolis: University of Minnesota Press.

Evans, B. and Reid, J. (2014) *Resilient Life: The art of living dangerously*. Cambridge: Polity.

Folke, C., Carpenter, S., Elmqvist, T., Gunderson, L., Holling, C. S. and Walker, B. (2002) 'Resilience and Sustainable Development: Building Adaptive Capacity in a World of Transformations', *AMBIO: A Journal of the Human Environment*, 31(5), 437–440.

Foucault, M. (1977) [1975]. *Discipline and Punish, the Birth of the Prison*. Harmondsworth: Penguin.

Foucault, M. (1980) *Power/Knowledge: Selected Interviews and Other Writings 1972–1977*. Brighton: The Harvester Press.

Foucault, M. (1988) 'Technologies of the Self', in Martin, L. H., Gutman, H., and Hutton, P. H. (eds.) *Technologies of the Self*. Amherst: University of Massachusetts Press.

Foucault, M. (1991) 'Governmentality', in Burchell, G., Gordon, C., and Miller, P. (eds.) *The Foucault Effect: Studies in Governmentality*. Chicago: University of Chicago Press.

Foucault, M. (1997) *Essential Works of Foucault 1954–1984*. Rabinow, P. (series ed.) Vol. 1 [Ethics, subjectivity and truth]. New York: The New Press.

Foucault, M. (1998) [1976] *The Will to Knowledge: History of Sexuality I*. Harmondsworth: Penguin.

Foucault, M. (2000) *Essential Works of Foucault 1954–1984*. Rabinow, P. (series ed.), Faubion, J. D. (ed.) Vol. 3 [Power]. New York: The New Press.

Foucault, M. (2003) *Society Must Be Defended, Lectures at the Collège de France 1975–1976*. New York: Picador.

Foucault, M. (2007) *Security, Territory, Population, Lectures at the Collège de France 1977–1978*. New York: Picador.

Foucault, M. (2008) *The Birth of Biopolitics, Lectures at the Collège de France 1978–1979*. New York: Palgrave Macmillan.

Foucault, M. (2010) *The Government of Self and Others: Lectures at the Collège de France 1982–1983*. New York: Palgrave Macmillan.

Foucault, M. (2011). *The Courage of Truth: The Government of Self and Others II. Lectures at the Collège de France 1983–1984*. New York: Palgrave Macmillan.

Gandy, M. (2004) 'Rethinking Urban Metabolism: Water, space and the modern city', *City*, 8(3), 363–379.

Gandy, M. (2006) 'Zones of Indistinction: Some thoughts on the bio-politics of urban space', *Cultural Geographies*, 13(4), 1–20.

Gausset, Q., Whyte, M. A., and Birch-Thomsen, T. (2005) *Beyond Territory and Scarcity: Exploring Conflicts over Natural Resource Management*. Stockholm: Nordiska Afrikainstitutet.

Greene, R. W. (1999) *Malthusian Worlds: U.S. leadership and the governing of the population crisis*. Oxford: Westview Press.

Halévy, E. (1995) *L'évolution de la doctrine utilitaire de 1789 à 1815*, Vol. 2. Paris: Presses Universitaires de France.

Hansson, S. (2015) 'Responsibilization in development cooperation', in S. Hansson, S. Hellberg, and M. Stern. (eds.), *Studying the Agency of Being Governed: Methodological reflections* (pp. 130–149). London: Routledge.

Hansson, S. and Hellberg, S. (2015) 'Power, freedom and the agency of being governed', in S. Hansson S., Hellberg, S., and Stern, M. (eds.) *Studying the Agency of Being Governed: Methodological reflections*. London: Routledge.

Hellberg, S. and Knutsson, B. (2016) 'Sustaining the life-chance divide? Education for sustainable development and the global biopolitical regime', *Critical Studies in Education*. DOI: http://dx.doi.org/10.1080/17508487.2016.1176064

Holden, E. and Linnerud, K. (2007) 'The Sustainable Development Area: Satisfying basic needs and safeguarding ecological sustainability', *Sustainable Development*, 15, 174–187.

Holling, C. S. (1973) 'Resilience and Stability of Ecological System', *Annual Review of Ecology and Systematics*, (4), 1–23.

Hornborg, A. (2003) 'Cornucopia of Zero-Sum Game? The Epistemology of Sustainability', *Journal of World-Systems Research*, 9(2), 3–14.

Joyce, P. (2003) *The Rule of Freedom: Liberalism and the Modern City*. London: Verso.

Kooy, M. and Bakker, K. (2008a) 'Splintered networks: Urban water governance in Jakarta', *Geoforum*, 39(6), 1843–1858.

Kooy, M. and Bakker, K. (2008b) 'Technologies of Government: Constituting subjectivities, spaces, and infrastructures in colonial and contemporary Jakarta', *International Journal of Urban and Regional Research*, 32(2), 375–391.

Lemke, T. (2002) 'Foucault, Governmentality, and Critique', *Rethinking Marxism*, 14(3).

Linton, J. (2010) *What is Water? The History of a Modern Abstraction*. Vancouver: UBS Press.

Lomborg, B. (2001) *The Skeptical Environmentalist*. Cambridge: Cambridge University Press.

Luke, T. W. (1995) 'On Environmentality: Geo-power and eco-knowledge in the discourses of contemporary environmentalism', *Cultural Critique*, No. 31, The Politics of Systems and Environments, Part II: 57–81.

Luke, T. W. (1999) 'Environmentality as Green Governmentality', in Darier, E. (ed.) *Discourses of the Environment*. Malden, MA: Blackwell Publishers.

Macy, J. (1991) *World as Lover, World as Self*. Berkeley: Parallax Press.

Malthus, T. (1798) *An Essay on the Principle of Population*. London: Dent & Sons.

Mayhew, S. (2004) 'Governmentality', in *A Dictionary of Geography*. Oxford: Oxford University Press.

McWhorter, L. (1992) 'Guilt as Management Technology: A Call to Heideggerian Reflection', in McWhorter, L. (ed.) *Heidegger and the Earth: Essays in Environmental Philisophy*. Kirksville, MO: Thomas Jefferson University Press.

Meadows, D. H., Meadows, G., Randers, J., and Behrens W. W. III. (1972) *The Limits to Growth*. New York: Universe Books.

Mikulak, M. (2013) *The Politics of the Pantry: Stories, food and social change*. Montreal: McGill-Queen's University Press.

Moseley, W. (1991) 'African evidence on the relation of poverty, time preference and the environment', *Ecological Economics*, 38, 317–326.

Neal, A. W. (2004) 'Cutting Off the King's Head: Foucault's Society Must Be Defended and the Problem of Sovereignty', *Alternatives*, (29), 2004.

Nightingale, A. (2011) 'Bounding difference: Intersectionality and the material production of gender, class and environment in Nepal', *Geoforum*, 42, 153–162.

Nightingale, A. and Ojha, H. R. (2013) 'Rethinking Power and Authority: Symbolic violence and subjectivity in Nepal's Terai Forests', *Development and Change*, 44(1), 29–51.

Ojakangas, M. (2005) 'Impossible Dialogue on Bio-Power', *Foucault Studies*, (2), 5–28.

Pirages, D. (2005) 'From Limits to Growth to Ecological Security', in Pirages, D. and Cousins, K. *From Resource Scarcity to Ecological Security*. Cambridge: The MIT Press, pp. 1–19.

Planet under Pressure. (2012) www.anthropocene.info/en/home [accessed 3 August 2014].

Plonowska Ziarek, E. (2012) 'Bare Life', in Sussman, H. (ed.) *Impasses of the Post-Global: Theory in the era of climate change*, Vol. 2. Ann Arbor: Open Humanities Press.

Reid, J. (2011) 'The Vulnerable Subject of Liberal War', *South Atlantic Quarterly*, 110(3), 770–779.

Reid, J. (2012) 'The disastrous and politically debased subject of resilience', *Development Dialogue*, No . 58, April 2012.

Reid, J. (2013) 'Interrogating the Neoliberal Biopolitics of the Sustainable Development–Resilience Nexus', *International Political Sociology*, 7(4), 353–367.

Rose, N. (1996) 'Governing "advanced" liberal democracies', in Barry, A., Osborne, T., and Rose, N. (eds.) *Foucault and Political Reason. Liberalism, neoliberalism and rationalities of government*. Chicago: University of Chicago Press.

Rose, N. and Miller, P. (1992) 'Political Power Beyond the State: Problematics of Government', *The British Journal of Sociology*, 40(2), 173–205.

Rutherford, P. (1999a) 'The Entry of Life into History', in Darier, E. (ed.) *Discourses of the Environment*. Malden, MA: Blackwell Publishers.

Rutherford, P. (1999b) 'Ecological Modernization and Environmental Risk', in Darier, E. (ed.) *Discourses of the Environment*. Malden, MA: Blackwell Publishers.

Rutherford, S. (2007) 'Green governmentality: insights and opportunities in the study of nature's rule', *Progress in Human Geography*, 31(3): 291–307.

Sandilands, C. (1999) 'Sex at the Limits', in Darier, E. (ed.) *Discourses of the Environment*. Malden, MA: Blackwell Publishers.

Scherr, S. J. (2000) 'A downward spiral? Research evidence on the relationship between poverty and natural resource degradation', *Food Policy*, 25, 479–498.

Skoglund, A., and Börjesson, M. (2014) 'Mobilizing "juvenocratic spaces" by the biopoliticization of children through sustainability', *Children's Geographies*, 12(4), 429–446. doi:10.1080/14733285.2013.824739.

Spivak, G. C. (1999) *A Critique of Postcolonial Reason: Toward a history of the vanishing present*. Cambridge, MA: Harvard University Press.

Spurr, D. (1994) *The Rhetoric of Empire: Colonial discourse in journalism, travel writing and imperial administration*. Durham: Duke University Press.

Stanescu, J. (2013) 'Beyond Biopolitics: Animal studies, factory farms, and the advent of deading life', *Phaenex. Journal of Existential and Phenomenological Theory and Culture*, 8(2), 135–160.

Sugiyama, K. (2015) 'Before the Law: Humans and other animals in a biopolitical frame by Cary Wolfe (review)', *SubStance*, 44(2), 176–180.

Sylvester, C. (2006) 'Bare Life as a Development/Postcolonial Problematic', *The Geographical Journal*, 172(1), 66–77.

Swyngedouw, E. (2010) 'Apocalypse Forever? Postpolitical Populism and the Spectre of Climate Change', *Theory, Culture and Society*, 27(2–3), 213–232.

Tellmann, U. (2013) 'Catastrophic Populations and the Fear of the Future: Malthus and the genealogy of liberal economy', *Theory, Culture and Society*, 30(2), 135–155.

Templeton, S. and Scherr, S. J. (1999) 'Effects of Demographic and Related Microeconomic Change on Land Quality in Hills and Mountains of Developing Countries', *World Development*, 27(6), 903–918.

UNEP (United Nations Environment Programme). (2004) *Exploring the Links: Human well-being, poverty and ecosystem services*. Nairobi: UN Publications.

UNEP (United Nations Environment Programme). (2007) GEO 4 (Global Environmental Outlook 4), www.unep.org/geo/GEO4/images/GEO4_cover_highres.jpg, [accessed 28 November 2014].

Venn, C. (2006) *The Postcolonial Challenge: Towards Alternative Worlds*. London: Sage.

Walker, J. and Cooper, M. (2011) 'Genealogies of Resilience: From systems ecology to the political economy of crisis adaptation', *Security Dialogue*, 14(2), 143–160.

Ward, L. (2013) 'Eco-governmentality Revisited: Mapping divergent subjectivities among Integrated Water Resources Management experts in Paraguay', *Geoforum*, 46, 91–102.

Watson, J. (2012) 'Butler's Biopolitics: Precarious community', *Theory and Event*, 15(2).

WCED (World Commission on Environment and Development). (1987) *Our Common Future*. Oxford: Oxford University Press.

Williams, R. (1976) *Keywords: A Vocabulary of Culture and Society*. London: Flaminga.

Wolfe, C. (2012) *Before the Law: Humans and other animals in a biopolitical frame*. Chicago: University of Chicago Press.

World Resources Institute. (2005) In collaboration with UNDP, UNEP, and World Bank. *Resources 2005 – The Wealth of the Poor: Managing ecosystems to fight poverty*. Washington, DC: WRI.

World Resources Institute. (2008) In collaboration with UNDP, UNEP, and World Bank. *World Resources 2008: Roots of Resilience – Growing the Wealth of the Poor*. Washington, DC: WRI.

Wrangel, C. (2014) 'Hope in a Time of Catastrophe? Resilience and future of bare life', *Resilience: International Policies, Practices and Discourses*, 2(3), 183–194.

Zevnik, A. (2009) 'Sovereign-less Subject and the Possibility of Resistance', *Millennium – Journal of International Studies*, 9 July.

3 Global water governance

Scarcity, populations, distinctions

Introduction

Chapter 2 outlined the theoretical approach of this book and a governmentality perspective on (neo)liberalism and power. Moreover, it provided a biopolitical perspective on the population–resource nexus in liberal economic government through problematizing the notion of scarcity, especially as it relates to different populations. One basic point is that, in this context, scarcity emerges as a 'regulatory device' (cf. Tellmann, 2013) that is made use of in order to govern people's use of resources. Furthermore, the chapter opened up to critical scrutiny the basic needs approach from a biopolitical perspective, arguing that it involves the drawing of lines between different forms of human life and between human lives and other forms of lives, as the approach is merely a means to satisfy human needs and aspirations. A critical perspective on the incorporation of the notion of resilience in a sustainable development regime was also introduced.

With this framework as a basis, this chapter analyses the governing rationales of global water governance (GWG) through a biopolitical lens. It critically studies GWG in relation to how it produces meaning around scarcity and how it portrays the role of different populations, with regard to the question of (basic) water needs, both of humans and ecosystems, and how it situates itself in relation to the notion of resilience. It also outlines how GWG can be understood as neoliberal in a Foucauldian sense. It should also be noted that because of the uncertainty of governing, as was emphasized in the previous chapter, together with the deficiencies that are specific to GWG, which are discussed below, we should be wary of assuming that the governing rationales of GWG, as presented in this chapter, are implemented or even implemented successfully.

The material consulted for analysing GWG in this chapter comes first and foremost from UN World Water Development Reports and other publications and sources connected to the UN's work on water, which is coordinated by UN Water. Second, it consists of publications from other central global water organizations, such as the Global Water Partnership (GWP). The third kind of organization whose material is considered are those with a broader mandate than water-related issues, but whose work is central to how the water agenda is developing on the global level, such as the World Health Organization (WHO). Before embarking on this biopolitical analysis an outline of some basic aspects of GWG is given, which includes a presentation of the overarching framework of Integrated Water Resources Management (IWRM), and the main criticism that has been directed against this approach, as well as other concepts and agendas that are central to GWG.

The first section briefly presents GWG – its main aims, actors, and arenas. It then outlines the new water governance discourse that emerged in the 1990s and discusses the Integrated Water Resources Management (IWRM) approach and how this framework has been criticized and developed over time. The concept of Water Security is then discussed as a central component of contemporary GWG, as is the role of water in the 2030 Agenda. In the second section of the chapter, the governing rationales of GWG are presented through a detailed biopolitical reading of central GWG publications and reports. The role and meaning of scarcity within GWG are discussed, as are the three main rationales around the productive uses of water for: (1) economic development; (2) basic human needs; and (3) ecosystems. The chapter finishes with a discussion of the notion of resilience within GWG and a conclusion.

Global water governance

GWG has earlier been described as 'the development and implementation of norms, principles, rules, incentives, informative tools, and infrastructure to promote a change in the behaviour of actors at the global level in the area of water governance' (Pahl-Wost *et al.*, 2008). In this book, references to the notion of GWG is based on a set of assumptions about the state of (global) water resources and (global) water problems, as well as solutions to these challenges. Hence, in line with the governmentality/biopolitics framework outlined in the previous chapter, GWG and its 'norms, principles and incentives' (ibid.) is approached as a particular way of understanding water issues that has productive effects in terms of establishing certain truths about water and proper ways of using and managing the resource. Such a way of defining water and its governance is developed in Chapter 4 in terms of how we can make sense of GWG as a 'neoliberal hydromentality'.

The core aim of global water governance, as expressed by UN Water, is working towards ensuring everyone's access to water (and sanitation) services while simultaneously creating 'a healthy, sustainable and prosperous world where no-one is left behind' (UN Water, 2017a). As conceptualized in this book, GWG is made up of global policy institutions, such as UN agencies, and of research institutions, donor agencies, NGOs, as well as global action networks such as GWP, The Gender and Water Alliance

(GWA), and The International Water Association (IWA).[1] Organizations with a panoply of issues over and above water but whose work is central to water issues, such as WHO, will also be understood as part of GWG, as indicated in the introduction to this chapter. These networks and organizations are tied together by the shared belief that water issues need a global policy response. This belief is grounded in the recognition that the challenges associated with water extend beyond national and regional boundaries and that they are too complex to be addressed by national and regional policies (Cooley *et al.*, 2013).

GWG has been developed in a number of state-led summits and forums, especially UN meetings around sustainable development, for example, the UN Conference on the Environment and Development in 1992, the UN Summit of 2000, the Rio+20 Summit in 2012, as well as the G8 meetings. Other important forums, which are pivotal for the development of GWG, include the World Water Forums and World Water Week (ibid.).

UN Water lists a number of water-related themes on their homepage, which is informative of the main themes within GWG. These themes are: climate change; disasters; ecosystems; financing; gender; human rights; quality and wastewater; scarcity; transboundary waters; urbanization; water, sanitation and hygiene; and water, food and energy (ibid.). I will come back to (most of) these themes throughout the chapter.

In order to tackle these issues, GWG emphasizes the need for international cooperation and includes both formal and informal mechanisms, such as creating norms, principles, rules, and incentives at the global level (Cooley *et al.*, 2013; Pahl-Wost *et al.*, 2008). Furthermore, GWG supports national and local water governance and sectors through: establishing international water agreements and laws; financing water resources management and service delivery efforts; establishing best practices; facilitating technology and knowledge transfer; conducting education and outreach programmes; and collecting, monitoring, and evaluating water-related data (Cooley *et al.*, 2013).

Despite the massive efforts that have been put into developing GWG by various organizations and networks it should be noted that its ability to be a stringent and influential global force is limited because of the lack of coordination between different agendas, agencies, and organizations (Cooley *et al.*, 2013; see also UN Water's *Strategy 2014–2020* [UN Water, 2014]). This lack of leadership, it has been noted, has consequences in relation to bilateral funding agencies as, because of this, they are more likely to push their own agendas and priorities in relation to water issues (Cooley *et al.*, 2013).

It is thus important to recognize that the actors, given their broad spectrum, diverge on a number of issues and that the stipulations of water governance on the global level are not necessarily successful in national and local contexts. In fact, many discussions in global water forums (still) revolve around how to go from policy to implementation and from principles to action. This has also been a central part of the critique that has been levelled at the IWRM framework, as is presented below. Despite the fact that GWG can be viewed as incoherent and fragmented, this book aims to paint a picture of some of its central agendas, themes, and policy prescriptions, starting with an outline of how the present paradigm of water governance emerged in the 1990s.

The emergence of a new paradigm

During the 1990s a new story about water emerged and paved the way for a new governance regime. Previous efforts to manage water resources were understood as 'inadequate' and 'misdirected' (Gleick, 1993, p. 3) and 'ignorant of the functioning of basic hydrological processes' (ibid.). According to this narrative, the world was, and still is, in the midst of a water crisis that is escalating because of increasing demand due to population growth, urbanization, industrialization, and an uncertain supply related to the pollution of water resources and climate change. This crisis narrative placed the *scarcity of water* at the centre of water management and, furthermore, problematized the unsustainable uses and management of water in terms of their negative effects on ecosystems, particularly on wetlands. At the same time there was recognition of the fact that many people still lacked adequate access to safe water. The ultimate effect of this development has been understood as involving increasing competition and conflict around water resources.

The governance shift that took place as a consequence of these new ways of understanding water challenges included a shift from centralized governance models, with an emphasis on technology and large infrastructure development as solutions to the problems of water, to decentralized systems with an emphasis on the *management* of water. Because of how public utilities were understood to have failed to provide universal access to water it also involved more focus on private actors, market solutions, and competition (Bakker, 2010). How this new understanding of water differed from the previous paradigm and how we can understand it in terms of a 'hydromentality' terms is explored in Chapter 4.

An important milestone for the development of the new paradigm was the Dublin conference on Water and the Environment in 1992. The conference called for 'fundamental new approaches to the assessment, development and management of freshwater resources' (ICWE, 1992), and it adopted four principles that have been guiding GWG for more than two decades. Those principles were embraced as important components of the World Bank water resources strategies (World Bank, 2004), as well as UN policies (see, for example, WWDR, 2003, 2006, 2009, 2012), and have been promoted by important international water organizations such as the GWP. The principles declare that:

1 Fresh water is a finite and vulnerable resource, essential to sustain life, development and the environment.
2 Water development and management should be based on a participatory approach, involving users, planners and policy-makers at all levels.
3 Women play a central part in the provision, management and safeguarding of water.
4 Water has an economic value in all its competing uses and should be recognized as an economic good.

(ICWE, 1992)

The first principle is understood as requiring an integrated and holistic approach that acknowledges water's role in many different purposes, functions, and services for both natural systems and human use. It also requires a holistic institutional approach that coordinates decision-making at different levels in relation to different uses of water. The second principle acknowledges that stakeholders at all levels should have an impact on water management decisions. The third principle involves recognizing the key role that women play in the collection and safeguarding of water for household purposes and that there is a need to increase women's participation in water decisions, as well as ensuring that the water sector is gender aware. The fourth principle means that the scarcity of water resources requires changing perceptions of water to an understanding in which its economic value is acknowledged. With regard to the economic dimension of water, which has been the most controversial in relation to the debate around what water *is* (in terms of framing it, for example, as an economic good or a human right, which is discussed more in detail in Chapter 4), the Dublin statement further declares that:

> Within this principle, it is vital to recognize first the basic right of all human beings to have access to clean water and sanitation at an affordable price. Past failure to recognize the economic value of water has led to wasteful and environmentally damaging uses of the resource. Managing water as an economic good is an important way of achieving efficient and equitable use, and of encouraging conservation and protection of water resources.
>
> (Ibid.)

A governance approach based on these criteria hence recognizes water as a right, as an economic good, and as a fundamental resource for the environment. These criteria, as well as the Dublin principles, are all embraced in the governing approach of Integrated Water Resources Management (IWRM), which grew out of the new ways of understanding water and which should tackle the challenges to water that had been identified. IWRM is commonly defined as 'the co-ordinated development and management of water, land

and related resources, in order to maximize the resultant economic and social welfare in an equitable manner without compromising the sustainability of vital ecosystems' (GWP, 2000a, p. 22). It is this framework – its initiatives and priority areas, and evolving concepts connected to this framework, such as water security (the relationship between IWRM and water security is discussed below) – that are addressed here as GWG.

Integrated Water Resources Management (IWRM)

The IWRM framework embraces three basic criteria for water management: social equity (ensuring access for all users to an adequate quantity and quality of water); economic efficiency (bringing the greatest benefit to the greatest number of users possible with the available financial and water resources); and ecological sustainability (ensuring the functioning of ecosystems) (see, for example, Lenton and Muller, 2009). The approach that derives from these criteria, known as the three E's, is thus closely connected to the mainstream definition of sustainable development in terms of its three pillars (environmental, social, economic).

With regard to national development of water governance, the IWRM framework emphasizes governance through its focus on: the *enabling environment* (national policies, and legislation and regulation of, and information to, the stakeholders); the *institutional roles* and functions performed by different stakeholders at different levels; and *management instruments* for regulation, monitoring, and enforcement (see the IWRM toolbox in GWP, 2017).

As stated in the Dublin principles, users, planners, and policy-makers should be involved in decisions regarding water management. This further includes the task of 'raising awareness of the importance of water among policy-makers and the general public' (ICWE, 1992). Additionally, '[i]t means that decisions are taken at the lowest appropriate level, with full public consultation and involvement of users in the planning and implementation of water projects' (ibid.). According to the IWRM approach, decisions in water management should not be left solely in the hands of the water experts. Ideally, this should guarantee a dialogue between different actors and interests. Participation is assumed to produce more effective, efficient, sustainable, and equitable resource management (Anokye and Gupta, 2012). Moreover, in accordance with the third Dublin principle, a gender-aware approach to participation is required, along with addressing 'women's specific needs' (ICWE, 1992). Women are to be empowered 'to participate at all levels in water resources programmes, including decision-making and implementation, in ways defined by them' (ibid.). Another central aspect of IWRM, which is related to participation, is knowledge production and information sharing. Within the IWRM approach, the sharing of knowledge and data between experts and stakeholders is understood as crucial for making informed decisions. Information and exchange are seen as 'a capacity building tool for the rest of the institutional set-up' (GWP, 2017). Four types of information are identified as important for supporting IWRM, these include: *data* on the characteristics of water resources, such as volumes and quality of water as well as spatial variabilities; *information* on how these data can be used for specific purposes; *knowledge* in terms of the implications of trends and values over time, and their impacts on resource use practices; and *wisdom* in terms of 'commonly accepted methods of using water resources to ensure sustainability' (ibid.).

For a broader understanding of IWRM, and for assessing the contribution that a biopolitical perspective makes to the literature on global water governance, it is important to note the massive debate that IWRM has stirred up in academia. The following is an outline of some of the main strands of criticism that have been directed against the framework, without claiming to be comprehensive. The criticism has been both of a general character and directed particularly against some of the focus areas of the Dublin principles, such as the way that the IWRM framework incorporates participation and approaches gender. Another main line of criticism is that IWRM is largely a technocratic approach that has a depoliticizing effect on how water issues are understood.

The more general critique that has been directed against IWRM is that the framework is vague and has failed to provide specific programmatic recommendations (Biswas, 2004; see also Jeffrey and Geary, 2006; Öjendal *et al.*, 2012; Varis *et al.*, 2008). Biswas argues: '[t]he absence of any usable and implementable definition has only compounded the vagueness of the concept, and has reduced its implementation potential to a minimum' (2004, p. 251). He further questions whether IWRM has improved water management practices anywhere in the world, and whether changes that have occurred are in fact due to the IWRM approach. Relatedly, it has been claimed that there is a 'mismatch between global prescriptions and local realities' (Chikozho, 2008, p. 1255). It has even been argued that IWRM is a 'nirvana concept' (Molle, 2008) that allows 'business as usual', as powerful actors appropriate the language of IWRM and use it to promote their agendas. Biswas writes:

> In fact, it appears that the vagueness of a concept, to a significant extent, increases its popularity, since people can then continue to do the same old stuff (SOS) they were doing before, but can concurrently claim that they are au currant with the latest global thinking. This jumping on the bandwagon also increases, often very significantly, the potential of receiving funding support from the donors, and also other personal benefits.
>
> (Biswas, 2004, p. 13)

Thus, this general critique that has been voiced against the IWRM approach has largely revolved around the vagueness of the approach and the so-called implantation gap (Cook and Spray, 2012). The framework is simultaneously accused of effecting no real change in terms of achieving a more socially, economically, and ecologically sustainable approach to water, but that it is merely a discourse without a proper connection to the realities on the ground. According to such a view, the IWRM does not become an alternative to the previous paradigm, but rather reinforces its roles, mandates, and worldviews (Molle, 2009, p. 135). Further, it has been argued that individual actors use the framework to gain benefits and attract funds, as indicated in the Biswas quote above, and that the inherent political aspects of water management, and the vested interests attached to it, hinder transformation (Biswas, 2004; see also, for example, Petit and Baron, 2009; Rahaman and Varis, 2005).

Another general concern, related to the above, has pointed to the enormous challenges of integrating the potential range of factors and the knowledge needed to manage water resources, according to the IWRM framework (Cook and Spray, 2012; Savenije and Van der Zaag, 2008). Regarding this debate, viewpoints differ in relation to whether a central problem of the IWRM approach is that it is dominated by engineers and physical science (see, for example, McDonnell, 2008), or rather, that the insufficiency lies in that the framework is not based on scientific principles (see Cook and Spray, 2012; Jeffrey and Gearey, 2006; Petit and Baron, 2009).

The literature on how participation has been performed in an IWRM framework (see, for example, Anokye and Gupta, 2012; Kujinga, 2002; Mostert, 2003; Njoh, 2002; Ozerol and Newig, 2008; Prokopy, 2005), has focused on factors that hamper effective participation. Such literature has elucidated, for example, how certain members of the public are excluded from participating in decisions regarding water because of a lack of proper representation in decision-making bodies and inadequate financial resources (Anokye and Gupta, 2012; Kujinga, 2002), Relatedly, it is also recognized that well-organized and resource-strong stakeholders, experts, and bureaucrats (and state representatives) tend to dominate decision-making processes and that '[h]ard-issues' of 'technology, finance and physical resources' are often prioritized (Njoh, 2002). Chapter 4 returns to the theme of participation. In this chapter, participation is problematized, from a governmentality perspective, in terms of how we can understand it as a governing technique employed to promote the interests of the (local) state.

The taken-for-granted narratives of IWRM in terms of the assumed links between water and gender, and the benefits of involving women in decision-making, have been interrogated by scholars focusing on the gendered nature of water (see Ahlers and Zwarteveen, 2009; Birkenholtz, 2013; Cleaver and Hamada, 2010; Harris, 2008, 2009; O'Reilly, 2006; Sultana, 2009, 2011). This research argues that what is critical is:

> the need to look beyond the mechanisms for accessing water at the local level to wider processes of governance, in order to understand how gender inequality is embedded and reinforced. This includes understanding how resources, actors, mechanisms and outcomes shape how women and other marginalised groups are able to gain access to water sources.
>
> (Cleaver and Hamada, 2010, p. 37)

Such a research endeavour, as described here, involves analysing the gendered effects of neoliberal governance reforms (see Ahlers and Zwarteveen, 2009; Harris, 2009), as well as how gender and gender–water relations are actively constructed, not only in the intersection between other social categories (such as class, sexuality, race, and ethnicity), but also in relation to livelihoods and access to land (see Harris, 2008) and to changes in the environment and physical landscape (Sultana, 2009).

Another strand of critique has argued that the IWRM approach involves an apolitical attitude towards water governance. Swyngedouw, for example, writes in his review of the 2012 World Water Development Report:

> A decidedly depoliticizing language is deployed that allows the report to introduce intensely political themes that nonetheless skirt around the key axes of struggle and conflict. For example, the commodification of water becomes veiled in a language of valuation and eco-services, with the refusal of the private sector to invest in the water sector as a problem of 'financing', social conflict as a question of stakeholder participation and proper institutions, egalitarian access to water and to the institutional mechanisms that organize its transformation and distribution as a problem of good governance. The political need to make clear choices and decisions, with implications often not fully accounted for, is camouflaged in a language of risk and uncertainty.
>
> (Swyngedouw, 2013, p. 826)

For Swyngedouw (as already quoted in Chapter 1), 'the water problem is not merely a question of management and technology, but rather, and perhaps in the first instance, a question of social power' (2004, p. 175). The ways in which critical research has politicized hydrosocial relations has been discussed in Chapter 1. Such research has included a focus on how meanings and representations of water and water problems are constructed in relation to political and social circumstances, and on the unequal distribution of both water and power, as well as on the productive effects of hydropolitical relations of power in terms of the creation of subjectivities. We will return to the political aspects of global water governance in the second half of this chapter.

More moderate reflection and critique within the paradigm itself have also emerged, for example, relating to the relationship between the different pillars (or criteria) of sustainable water management (social equity, economic efficiency, and ecological sustainability). Much emphasis has been placed on the economic aspects of water at the expense of other dimensions. Economic valuation of water has been argued as critical for improving water services; charges need to be widely implemented to improve cost recovery, to maintain and expand infrastructure, and to provide incentives for conservation and reuse. The 2015 World Water Development Report (WWDR, 2015) acknowledges, however, that while much work within the IWRM framework has been geared towards economic efficiency, there is a need to put more emphasis on issues related to both social equality and environmental sustainability (ibid., p. 6). It states (building on WGF, 2012) that even though the IWRM framework is steered by the concern for all three dimensions of sustainable development, 'the social equity goal is often given less priority when water allocations decisions are made' (WWDR, 2015, p. 21). Here, it is also recognized that the limited realization of water as a human right has unequal effects on the poor, and on women and children in particular (ibid., p. 22).

This debate is indicative of the fact that the frameworks of GWG are not static and that, following some of the critique mentioned above, they have changed significantly over the years. Even though there still remain challenges in this regard, IWRM in its most technocratic and managerial form has had to give way to an approach that has increasingly taken social aspects of water systems into account (Srinivasan *et al.*, 2017). Other frameworks and concepts in relation to water issues, such as water security and the nexus approaches (food–water–energy nexuses) have also emerged. However, IWRM, in a broad sense, is still the overarching framework for GWG (see, for example, Benson *et al.*, 2015). For instance, IWRM as a concept is explicitly referred to in the sustainable development goals (SDGs); target 6.5 sets out to, '[b]y 2030, implement integrated water resources management at all levels, including through transboundary cooperation as appropriate' (UN, SDG 6, 2017).

The concept of Water Security

Among new emergent concepts, water security has become one of the most influential in global water policy discussions during the last decade. However, academics, policy institutions, and organizations apply multiple definitions of the concept. Therefore, no single definition of the term exists. Cook and Bakker (2012) have identified four main themes that dominate contemporary approaches to water security: *water availability*; *human vulnerability to hazards*; *human needs*; and *sustainability*. As we shall see, contemporary approaches embrace a broad understanding and thereby include most, if not all, of these themes in their definitions. Effectively, the term has had little reference to military security and so-called water wars, a theme that appeared in the 1990s and involved the stipulation that scarcity of water would lead to conflict between states over the resource (ibid.).

Early understandings of water security were, however, human centred (Srinivasan *et al.*, 2017). Human-centred definitions typically focus on the individual level and define water security as access to sufficient and affordable water to satisfy human needs. Such an understanding of water security is close to how the right to water has been defined (see, for example, Rijsberman's definition in Cook and Bakker, 2012, p. 97), which is discussed further on in this chapter. What recent approaches have in common is that they focus on both human and ecosystem needs (ibid.). One example of such a broad definition, which is based on UNESCO's International Hydraulic Programme, is proposed in UN Water's analytical brief:

> Water security is defined here as the capacity of a population to safeguard sustainable access to adequate quantities of acceptable quality water for sustaining livelihoods, human well-being, and socio-economic development, for ensuring protection against water-borne pollution and water-related disasters, and for preserving ecosystems in a climate of peace and political stability.
>
> (United Nations University, 2013, p. 1)

Another common definition by the GWP reads as follows:

> Water security, at any level from the household to the global, means that every person has access to enough safe water at affordable cost to lead a clean, healthy and productive life, while ensuring that the natural environment is protected and enhanced.
>
> (GWP, 2000b, p. 1)

The 2015 WWDR presents the following definition of water security:

> Water security: The availability of an acceptable quantity and quality of water for health, livelihoods, ecosystems and production, coupled with an acceptable level of water-related risks to people, environment and economies.
>
> (Grey and Sadoff, 2007, cited in WWDR, 2015, p. 86)

These definitions make it apparent that the notion of water security has many similarities with the IWRM framework. When understood in broad terms, the concept overlaps with IWRM, as it is a paradigmatic approach that integrates different scales, balances human and ecosystem needs, and includes questions of both quality and quantity (Cook and Bakker, 2012, p. 98). Such a broad and complex understanding of water issues creates challenges for implementation. This means that the critique directed at IWRM for being a merely theoretical approach that is difficult to transform into practice, is also relevant in relation to the concept of water security (ibid.).

One difference between the two approaches is, however, that compared to IWRM, water security offers a vision for a 'desired end state', (that is, water security) (Mitchell, 2006, p. 53, cited in Cook and Bakker, 2012, p. 98) and that it implies 'thresholds' beyond which water is not secure (Cook and Bakker, 2012, p. 98). Additionally, framing something in terms of security can also be a way of placing an issue higher up the list of priorities. In that sense, water security might have advantages, compared to the IWRM approach, in driving the global water agenda. Other distinct features of water security are that it is more focused on risk, threats, and vulnerabilities, presented, for example, by scarcity, pollution, and flood and drought, not least in connection to climate change. In relation to this theme, even if it is rarely discussed in such terms, the question of state security aligns with the water security concept because of the state's need to protect crucial water infrastructure. Both narrower and broader definitions place the issue of the human need for water at the centre of water security. Defining the amount of water required for basic needs, as well as enforcing and evaluating the progress of realizing the human right to water, are important features of water security, and of GWG in general. We return to the question of the right to water and the theme of risk and vulnerabilities, and how we can understand this biopolitically, at a later stage in this chapter.

Water and Agenda 2030

Water issues and governance are part of Agenda 2030 for Sustainable Development. While the Millennium Development Goals (MDG) water target 7C, which focused on halving 'by 2015, the proportion of the population without sustainable access to safe drinking water', was met globally five years ahead of schedule (UN, 2017a), the SDGs are even more ambitious. The SDG target 6.1 sets out to '[b]y 2030, achieve universal and equitable access to safe and affordable drinking water for all' (UN, 2017b). This means that an additional 663 million people need to be provided with improved drinking water sources (WHO and UNICEF, 2015: 4).

The SDG water goal is also ambitious in the sense of setting broader targets in relation to: water quality (6.3); water use efficiency and scarcity (6.4); implementing integrated water resources management including transboundary cooperation (6.5); protecting and restoring water-related ecosystems (6.6); expanding international cooperation and capacity building in water- and sanitation-related activities in developing countries (6.7); and supporting and strengthening local participation in water and sanitation management (6.8).

The experiences from the results of the MDGs have informed the formulation of the SDGs. The formulation in MDG water target 7C requested 'sustainable access' to 'safe drinking water'. 'Safe water' was measured against the indicator 'improved' water sources, in turn understood as sources protected against faecal contamination. However, to address criteria such as accessibility, availability, and quality, a consensus regarding the need for improving this indicator has been established. Therefore, the indicator for the SDG target 6.1 'safely *managed* drinking water services' was proposed to address this (UNSD, 2016, p. 21, my italics).

This means that while the MDG target had a narrower focus on access to water (and sanitation), the indicator for the SDG 6.1 also takes into account the maintenance of water systems. Hence, this is a call for the water community to rectify the problem that, in the course of achieving the MDGs, not enough funding was allocated to the operation and maintenance costs of water supply systems (Cooley *et al.*, 2013, p. 17). South Africa is a clear-cut example of the different challenges involved between MDG water target 7C and the SDG target 6.1. While the country was doing very well in terms of extending water (and sanitation) services to new areas and users, it now faces vast challenges in terms of maintaining its degrading infrastructure.

Because of the complexity of these targets, it has been acknowledged that there will be challenges in terms of measuring and tracking SDG 6 (Srinivasan *et al.*, 2017). The following focuses mostly on issues that are connected to target 6.1, as this target directly relates to the discussion on the right to (basic) water and how to define and quantify such a right; it also touches on issues around target 6.4 on water use efficiency and scarcity.

Governing rationalities of global water governance

This section focuses on the overarching governing rationalities of GWG and how they relate to the analytical themes developed in the previous chapter. It analyses the role of scarcity in GWG, and looks in more detail into the relationship between the economic, social, and ecological components of water governance. It also discusses the role that the notion of resilience plays in GWG.

Scarcity and (a particular form of) governance

Scarcity is a central problem of GWG. In water governance contexts, this scarcity is often illustrated by the share of water available for human consumption. Typically, this is described as only 0.3 per cent of the water on planet earth is usable by humans. The message that is conveyed in such accounts is that the problem around water is related to the physical availability of the resource and, furthermore, that we are in a situation where there is always too little water.

Scarcity has a peculiar role in contemporary global water discussion, however. While the scarcity of water was the starting point for the emergence of the global water crisis narrative, one of the basic messages of the IRWM framework is that the water problems we face today, rather than being merely a product of limited supply, are problems of *governance*. It is thus understood as both a human and a physical problem. WWDR 2012 states, for example:

> Water institutions are still largely technology and water supply driven. To improve the effectiveness of these institutions, the emphasis has to gradually change from technological solutions to *management of processes and people*, involving inclusive decision-making and bottom-up approaches.
>
> (WWDR, 2012, p. 7, my italics)

Similarly, the 2015 WWDR defines development practices and governance failures as causes of the increasing pressure on freshwater sources:

> Unsustainable development pathways and governance failures have generated immense pressures on water resources, affecting its quality and availability, and in turn compromising its ability to generate social and economic benefits. The planet's capacity to sustain the growing demands for freshwater is being challenged, and there can be no sustainable development unless the balance between demand and supply is restored.
>
> (WWDR, 2015, p. 10)

Different notions and measurements of water scarcity are used to describe the state of freshwater resources globally. One way of expressing the differences between different types of scarcity is to distinguish between physical, economic, and institutional water scarcity. *Physical scarcity* is when water is limited in quantity or quality. Globally, it is estimated that about 700 million people in 43 countries are presently in a state of water stress or scarcity. This is predicted to extend to more than 2.8 billion people in 48 countries by 2025 (GWI, 2013, p. 3).[2] In contrast, *economic scarcity* depicts a situation where lack of infrastructure is the reason that water does not reach users, while *institutional scarcity* describes a situation where the lack of capacity and institutional framework is the reason that reliable access to water cannot be ensured (FAO, 2016). In 2007 it was estimated that about 1.8 billion people suffered from economic water scarcity (UN Water, 2007).[3]

In terms of physical water scarcity, countries are commonly divided into the following groups: water-stressed countries (when the annual water supply drops below 1,700 m^3 per person); water scarcity (when the annual water supply drops below 1,000 m^3 per person); and absolute scarcity (when the annual water supply drops below 500 m^3 per person). This method, however, fails to take into account different water situations in terms of the specific patterns of usage and solutions in different countries to deal with availability (cf. FAO, 2012, referenced in WWDR, 2016). Therefore, another measure focusing on the human pressure on water resources has been proposed. This measure is 'based on the ratio between water withdrawal by agriculture, municipalities, and industries over total renewable water resources' (WWDR, 2016, p. 16, building on UNSD, n.d.).

This was how MDG indicator 7.5 on the 'proportion of total water resources used' was measured (UN, 2008, p. 2). SDG indicator 6.4.2 – 'Level of water stress: freshwater withdrawal as a proportion of available freshwater resources' – builds on this measure and also takes into account environmental water requirements (UNSD, 2016, p. 19). These environmental water requirements 'are established in order to protect the basic environmental services of freshwater ecosystems' (ibid.). In the metadata for the SDGs, it is, however, acknowledged that there are no universally accepted thresholds for assessing the sustainability of water withdrawals (ibid., p. 21). To classify levels of water stress, three categories are proposed: 'low, high and very high' (ibid., p. 20). It is further stated that the thresholds 'could be country specific, to reflect differences in climate and national water management objectives' or that 'uniform thresholds could be proposed using existing literature and taking into account environmental water requirements' (ibid.). However, there has been critique against such ways of measuring water use pressure. Focusing on data on the national level masks both in-country variability and the fact that water is transboundary. Thus, water stress levels can vary within large countries and be shared by two or more nations. To address the latter, basin level analysis is proposed (WWDR, 2016, p. 17).

In conclusion, there is no single approach to *defining* and *measuring* water scarcity within the frames of GWG, and the productive role that scarcity plays is complex. Thus, scarcity, as it is understood here, has something of an 'uncertain status' (Tellmann, 2013). Nevertheless, through the way that the *idea* of water scarcity is framed in policy descriptions, it emerges as a 'regulatory device' (ibid., p. 137) that is meant to create a sense of futurity, and a reorientation towards acting ' "as if" scarcity prevails in any given moment' (ibid., p. 146). For example, WWDR 2015 (p. 11) predicts (based on 2030 WRG, 2009) that unless new ways of managing and using water are implemented, a 40 per cent global water deficit will be experienced by 2030. Perhaps this is expressed even clearer on UN Water's homepage, where it is stated that:

> There is not a global water shortage as such, but individual countries and regions need to urgently tackle the critical problems presented by water stress. Water has to be treated as a scarce resource, with a far stronger focus on managing demand.
>
> (UN Water, 2017b)

Demand management has, since the establishment of IWRM as the dominant approach to water governance, been seen as the proper way of dealing with water availability. As noted in the first section of this chapter, a focus on demand rather than the production of new water supplies through large-scale infrastructural investments, such as dams, has been a main message of this framework. At present, however, while there is an acknowledgement of the complexity of water scarcity, in that it is a combination of human factors and physical availability (as well as problems in measuring it), there is a tendency in GWG to focus on physical water scarcity. For example, in the latest WWDR (2017), there is no mention of economic (or institutional) water scarcity but a complete focus on the physical availability of water. This report focuses on waste water treatment as a way not only of preventing pollution due to untreated water released into the environment without purification, but also of the possible reuse of cleaned water. Together with desalination, waste water treatment is at the forefront of how to deal with water shortages. Such solutions for water scarcities are well in line with overarching rationales of sustainable development in terms of the focus on technological solutions to problems of the environment and scarcity, rather than focusing on their underlying causes. In a water context, however, this might be a game changer as this development can be seen as a return to supply-side management and a (re)turn to infrastructure (Crow-Miller *et al.*, 2017) as solutions to water availability. Thus, it remains unclear at this particular time how these dynamics will affect GWG in the ways it addresses water scarcity in terms of whether it is mainly a population's water *use* or water *availability* that should be amended and managed. In practice, we will certainly see a combination of both.

We can conclude that the predominant focus on measurement of scarcity at national or regional levels obscures distributional issues and the power relations that determine which populations and sectors get how much water, when, and how. We will also see further on in this chapter how GWG is haunted by Malthusian logics with respect to how different populations are understood regarding their relation to ecosystem services and the consequences following from ecosystem decline, in relation to which water scarcity (as well as quality) is central.

What is especially important from the perspective of this book is, however, how the notion of scarcity is used in strategies of governing in particular contexts. Ultimately, this is an empirical question, since the specific mentalities and the techniques used to govern water use and water users vary from context to context – a theme returned to in both Chapter 4 on hydromentalities and Chapter 5 on the effects on the particular hydromentality of eThekwini.

One basic point here is that scarcity within GWG is constructed in relation to the needs of the resource in the current political and economic systems (see also Linton, 2010; Loftus, 2015; Swyngedouw, 2015). This means that we need enough water to uphold the engines of growth and economic development, at the same time as meeting the needs of humans and ecosystems. This constant striving towards 'more' and 'better' means, in turn, that water will always be seen as 'scarce'. We now turn to one of the main governing rationales of GWG, namely, water for economic development.

Water for economic development

A cornerstone of both sustainable development and GWG is the importance of economic development. The 2016 WWDR, which focused on 'Jobs', stated that: 'Water scarcity is likely to limit opportunities for economic growth and the creation of decent jobs in the upcoming years and decades' (WWDR, 2016 p. 23). Thus, scarcity emerges here as a threat to socio-economic development. While the previous WWDR 2015 had stated, in line with both the Dublin principles and a sustainable development discourse, that:

> Economic development and water are intimately connected in many ways. Water is an essential resource for economic production and an 'enabler' of trade for most types of goods and services. Water is an essential input for the production of food and electricity, as well as for many manufactured products. Investments in water infrastructure are therefore fundamental to unlocking the full potential of economic growth.
>
> (WWDR, 2015, p. 23)

Hence, GWG, just as sustainable development in general, adheres strongly to an economic growth paradigm. In contemporary sustainable development discourse, this paradigm is described in terms of the 'green economy'. A green economy is an economy that 'results in improved human well-being and social equity, while significantly reducing environmental risks and ecological scarcities' (WWDR, 2015, p. 28, building on UNEP). Thus, a green economy 'can be thought of as one which is low carbon, resource efficient and socially inclusive' (ibid.). Sustainable development discourse and GWG consequently remain in the belief that environmental and resource concerns, as well as questions of equity, can be dealt with, and 'make way for a new era of economic growth' (cf. WCED, 1987, part 1, paragraph 27), within the frames of the current economic system. At the same time, however, there is a recognition that economic development may have negative impacts on ecosystems and the quality and quantity of water available (WWDR, 2015, pp. 2, 10, 30, 97).

Moreover, we can also recognize, in GWG, the rationalities surrounding poverty in more general sustainable development discourse. WWDR 2015 stipulated that:

> The relation between water and poverty is a two-way street. Poverty itself can have negative effects on the management of water resources and services. *The desperation and limitations arising from poverty can be a driver of pollution and unsustainable use of water resources.* Poverty can also render existing investments in water less efficient, since households and communities often find it difficult to finance, operate and maintain infrastructure such as rural water pumps. This poses a serious threat to long-term development and poverty reduction.
>
> (WWDR, 2015, p. 20, my italics)

Even if the language surrounding poverty, and its relationship to the ability to care for the environment, has softened over the last decades, this quotation bears reference (see the text in italics) to the 'downward spiral' and to the idea that the poor 'operate with a higher rate of time preference' (Moseley, 1991) compared to wealthier individuals. According to this understanding, it is assumed that the poor are preoccupied with surviving the present rather than saving for the future and thus exploit the environment, as discussed in the previous chapter. There is no doubt that such understandings inform the way that poor people's access to and use of water are governed and understood, both because of their ability to pay but also because of an ethical stance in relation to who can be understood as a responsible and sustainable water user. We come back to this in the next two chapters.

At the same time as adhering to an economic growth paradigm, symptomatic of the current global political economy, there is a recognition that the way that water resources are allocated is 'deeply problematic' and 'typically inequitable' (WWDR, 2015, p. 21). This, in turn, calls for mechanisms that take into account 'the interests of poor people, and that can mediate grievances between different users' (ibid.). The right to (basic) water is a strong discourse in GWG in order to come to terms with inequitable access.

Water for basic human needs

In 2010 safe and clean drinking water was explicitly recognized by the UN assembly as a human right (see UN, 2010). This recognition builds on a number of previous UN resolutions, covenants, declarations, and reports related to water as a human right, including Comment No. 15 of the Committee on Economic, Social and Cultural Rights. Article I.1 of this Comment states: 'The human right to water is indispensable for leading a life in human dignity. It is a prerequisite for the realization of other human rights.' Comment No. 15 further defined that the right to water 'entitles everyone to sufficient, safe, acceptable, physically accessible and affordable water for personal and domestic uses' (OHCHR, 2010, p. 1).

'Sufficient' water is understood here as the amount of water needed for personal and domestic use, such as drinking, sanitation, washing, food preparation, and hygiene (UN Water, 2017c). WHO has recommended a figure of between 50 and 100 litres per capita per day (lcpd) to fulfil these needs. However, figures as low as 20 lcpd are sometimes used, but such a low amount might raise health concerns, it is noted (OHCHR, 2010, p. 8). 'Safe' refers to the fact that the water needs to be free from contamination (UN Water, 2017c). 'Acceptable' means that the water should be acceptable in terms of odour and taste, and the way it is accessed should be acceptable in relation to culture, as well as being 'sensitive to gender, life-cycle and privacy requirements' (ibid.). 'Physically accessible' refers to the fact that the water access point should be within or in the close vicinity of the household, according to WHO not more than 1,000 metres, or 30 minutes, away from the household (ibid.). 'Affordable' means that everyone should be able to afford water and water services. UNDP suggests that the cost of water should not exceed 3 per cent of household income (ibid.).

Such a focus on water for basic needs is incorporated into the MDGs and the SDGs. There is evidence of progress for this agenda; the water dimension of MDG target 7C, which focused on halving 'by 2015, the proportion of the population without sustainable access to safe drinking water', was met five years ahead of schedule (UN, 2017a).

Despite such success, and despite the fact that the SDG water targets are even more ambitious than the MDGs, there are reasons for being critical of these global initiatives. Both MDG and SDG water targets for individuals represent a narrow water agenda that merely focuses on providing *drinking* water for all. It has been argued that such a focus on safe and clean water for domestic use is limiting in terms of the impact on people's livelihoods, since it does not recognize the importance of water for productive use, such as subsistence farming (Hall *et al.*, 2014; see also van Koppen *et al.*, 2009). Such a view is also supported by the 2015 WWDR, which, under the rubric of social aspects of water, acknowledges the importance of water not only for health but also for social dignity and for livelihood opportunities, such as agriculture and family businesses.

From a biopolitical perspective there is even more at stake. Defining the appropriate amount of basic water involves determining the amount of water required for essential needs. In a neoliberal framework, drawing this line means that above the level of survival, water will be the target of commercialization and marketization. This in turn implies that a line is drawn between those who are expected to rely on water merely for basic needs and those who can access larger amounts, if they have the means to pay for it. Hence, one of the main arguments of this book is that the basic water agenda involves a biopolitical distinction between different forms of lives (see also Hellberg, 2014, 2017). Such a separation – between those who should be content with satisfying basic needs and those who have unlimited access to water – can be understood as a central feature of neoliberal ways of governing. As was outlined in the previous chapter, neoliberalism, as an art of governing, acknowledges the need to work on the *conditions* that allow the market to function (Gudmand-Høyer and Lopdrup Hjorth, 2009, p. 118). From such a perspective, basic services can be seen as instruments that work to *provide a minimum level of security* so that market principles can be allowed to function in the rest of society (Foucault, 2008, pp. 206–207).

Basic water can hence be regarded as a policy instrument that *supports* market principles rather than being their counterpoint, as it 'allows the economic mechanisms of the game' (ibid., p. 206) to function above the threshold of survival. Bakker has argued, likewise, that defining water as a right does constitute a radical alternative to those who oppose the commercialization and privatization of water (see Bakker, 2010, 2007). This means that the particular form of governance adopted by the IWRM approach is, from a Foucauldian perspective, characteristic of neoliberalism and not to be seen as an alternative to such governance.

As the following two chapters examine further, South Africa has implemented a policy of FBW of 25 lcpd as a consequence of the South African constitution having included a right of access to sufficient water (RSA, 1996, 27[1][b]). This figure of 25 lcpd has been under debate, however, not least since the *amount* of basic water, as well as *the way this water is delivered*, have been an issue for the constitutional court. Chapter 5 empirically explores how this separation simultaneously produces and enhances difference in terms of the lifestyles that can be led, depending on what population you belong to.

From a biopolitical perspective, it is also important to acknowledge that even if the *human* right to water has now been recognized by the UN assembly, it does not mean that every human can or will enjoy such a right. As has been pointed out already by Hannah Arendt, the idea of 'universal human rights' is 'artificial' (Arendt, 1998, p. 2, cited in Wolfe, 2012, p. 6) because they are grounded in a biological notion of humans as *homo sapiens*; while in practice they are 'a product of membership in a political community' (Wolfe, 2012, p. 6). Effectively, there are individuals and populations, especially stateless people and migrants, and populations in war zones and failed states, who are left out of such biopolitical strategies of governing. This means that people will be forced to live without access to safe water, not only because of a lack of resources or implementation failures, but also because of the way that the system of rights is constructed. This aspect of the biopolitics of water is acknowledged in this book. However, its preoccupation is with governing strategies that do try to pin down the relationship between (different) populations and (water) resources and to solve distributional issues according to particular governing logics guided by sustainable development discourse.

In line with this task, we now turn to the third central governing rationality of GWG, namely, water for ecosystems. This theme is indicative of a separation between human life and other forms of life central to Western political thought in general and that of sustainable development and GWG in particular.

Water for ecosystems

While there is a recognition that '[a]quatic ecosystems are at the centre of all life' (WWDR, 2015, p. 28), the focus on ecosystems in GWG tends to remain on ecosystem *services* and ecosystem-based *management* (EBM). WWDR 2015 draws on Millennium Ecosystem Assessment (MEA, 2005), when describing ecosystem services in terms of four main categories for ecosystem services: provisioning (e.g. clean water); regulating (e.g. flow regulation and flood control); cultural (e.g. recreation); and supporting (e.g. habitat for aquatic species) (WWDR, 2015). UN Water's homepage declares that:

> A paradigm shift is taking place, with ecosystems being recognized as an integral part of development solutions. This reflects the steps being taken towards better integrated water resources management, and therefore more sustainable development.
>
> (UN Water, 2017d)

In turn, EBM is defined, based on the United Nations Convention on Biological Diversity (CBD), as a 'strategy for the integrated management of land, water and living resources that promotes conservation and sustainable use in an equitable way' (CBD, 2014). In this context, the notion of 'natural infrastructure' is important, understood as interconnected networks of ecosystems which both maintain those ecosystems and provide some of the same services as does human-built infrastructure. Because natural infrastructure provides such services, it is recognized that it involves economic benefits (WWDR, 2015, p. 29) and the challenge, as it is formulated in WWDR 2015, is to 'maintain a beneficial mix between built and natural infrastructure and provision of their respective services' (ibid., p. 30).

One of the main messages in relation to ecosystems is that ecosystem services are 'under-valued, under-recognized and under-utilized within economic and resource management approaches' (ibid., p. 29). WWDR 2015 therefore calls for a more 'holistic focus on ecosystems for water and development' (ibid.) in order not to worsen the difficulties of balancing water supply between different users and uses (ibid.). Compared to the MDG framework, which did not incorporate such a holistic stance towards water, an EBM approach is predicted to play an important role in addressing SDG targets (ibid.), which are much broader in their scope in relation to water issues, as has been outlined above.

Hence, the focus on ecosystems, as presented within GWG, is strongly anthropocentric, since well-functioning ecosystems attain their value by means of how they contribute to human development and economic benefits. Viewing nature and other forms of life as a resource base for human needs and well-being becomes an even stronger discourse in contemporary sustainable development politics, compared to the Brundtland report, when described through the language of ecosystem services and EBM.

However, at the same time, GWG texts show an apprehensiveness towards economic growth. In the context of ecosystem preservation and functioning and the services they provide, both population growth and economic growth are recognized as a strain (WWDR, 2015). Along with such human-produced drivers, climate change is also expected to have severe effects on ecosystem services and biodiversity (Boelee, 2011, referenced in WWDR, 2015). Moreover, poor water management is seen as a driver of the decline in biodiversity health (WWDR, 2015; WWF, 2012). That the current economic development paradigm cannot be questioned, however, despite its negative impact on ecosystems, is well illustrated in the following quote:

> While economic development may lead to ecosystem decline, ecosystem services underpin economic development, so the real challenge is in building awareness of the economic value of healthy ecosystems.
>
> (WWDR, 2015, p. 30)

Such economic arguments are recognized as a way of making ecosystem preservation relevant to decision-makers (ibid.). In addition, such an instrumental stance towards economic arguments as guiding principles in water and ecosystems governance makes *ecosystem valuation* a primary point of departure in the pursuit of sustainability. Value estimates of ecosystem services abound:

> According to the study 'Changes in the Global Value of Ecosystem Services', the 2011 economic value of ecosystems has been globally estimated at US$124.8 trillion. Global GDP was estimated at US$75.2 trillion in the same year.
>
> (Costanza *et al.*, 2014, cited in WWDR, 2015, p. 31)

On UN Water's homepage, the same study is referenced:

> Ecosystems across the world, particularly wetlands, are in decline in terms of the services they provide. Between US$4.3 and US$20.2 trillion per year worth of ecosystem services were lost between 1997 and 2011 due to land use change.
>
> (Costanza *et al.*, 2014, cited by UN Water, 2017d)

Ecosystem valuation is based on what users would be willing to pay for the services provided, or what it would cost to replace the services that ecosystems provide with human-built infrastructure (WWDR, 2015, based on Boelee, 2011). This approach is seen as the way forward to build a green economy in the post-2015 agenda (WWDR, 2015, p. 31).

Biopolitical logics come to the fore in connection with how different populations are understood in relation to ecosystem degradation. Degrading ecosystems are projected to have most effects on the poorest in terms of their food and water insecurity. Such effects are associated with the risk of 'resource conflicts' (WWDR, 2015, p. 30). WWDR 2015 states that:

> As populations increase and ecosystem services decline, the risk of resource conflicts rises especially where tensions already exist along ethnic or socio-economic lines.... Ecosystem degradation and climate change have significant potential to increase these tensions.
>
> (Ibid.)

Hence, not only is water and food security at risk for certain populations (the poor) as a consequence of ecosystem degradation, but so is their security understood more traditionally as related to the threat of violence and violent conflict. Again, we see Malthusian logics at play. The imbalance between resources (now called services) and population, implies a looming crisis for those who are unfortunate enough to have to rely on local resources for their livelihood, as well as for the region where they live.

Such a statement invokes the debate on the links between environmental degradation and violent conflict that emerged in the 1990s (Homer-Dixon, 1994, 1999; Kaplan, 1994), and that has been ongoing between scholars applying a neo-Malthusian logic focusing on scarcity and their critics (see, for example, Giordano *et al.*, 2005; Theisen, 2008)

In the water context, this debate was initially centred around the notion of 'water wars' (Starr, 1991) between states. The hypothesis of the emergence of such water wars as a direct consequence of water scarcity has been studied, and refuted (Wolf *et al.*, 2003). Even though more recent studies have focused on a local level when examining the relationship between scarcity and conflict, a causal link between the two has not been confirmed (see Mildner *et al.*, 2011, for an overview of this literature). Instead, researchers focusing on the role of resources for economy and society have pointed to the importance of governance and distributional issues in determining the outcomes for particular societies and, hence, in determining whether conflicts escalate or not (Benjaminsen and Ba, 2009; Moyo, 2005; Saad-Filho and Weeks, 2013). Nevertheless, postulations of such links, as in the quotation above, keep re-emerging.

The statement regarding the connection between population growth, the decline of ecosystem services, and conflict, also invokes the, so-called, security–development nexus. Many different ways of understanding the links between the two concepts exist (Stern and Öjendal, 2010). However, the nexus suggests that development and security are 'inextricably linked' and that poverty and lack of resources, such as water, provide a 'breeding ground' for conflict (see, for example, Kofi Annan, cited in United Nations, 2004: vii; DFID, 2005: 3, both cited in Stern and Öjendal, 2010, p. 5). From a biopolitical perspective, one of the most important critiques of the security–development nexus has been that, as it has been imbued with meaning in the policy world, it is a way of keeping dangerous life at bay through biopolitical regulation (Buur *et al.*, 2007; Duffield, 2007). Such critique of biopolitical regulation relates to the more recent incorporation of the concept of resilience into environmental and resource governance discourse, resulting in the constitution of a 'Sustainable Development–Resilience Nexus' (Reid, 2013). How the concept of resilience has been discussed and applied in the water context is presented below.

Risk, uncertainty, and resilience

As should be clear by now, GWG operates with a specific understanding of the world and the role of water in it. Furthermore, as was outlined in the previous chapter, the perception that our existence in the world is characterized by risk and uncertainty has recently grown within the frame of neoliberal governance and in relation to environmental questions. This is also the case in relation to water governance. Hence, the concept of resilience has been increasingly employed in water management discourse in the last few years (see, for example, Bohensky, 2008; Rockström *et al.*, 2014; WHO, 2009).

The main point of such an approach is that an 'adaptive capacity' (WHO, 2009) in water management is needed, especially considering the unknown effects of climate change. Climate change is an increasingly important factor in predicting, and planning for, the accessibility of water resources. Both floods and droughts are likely to become more frequent with climate change. UN Water (2010, p. 9) states:

> Owing to future uncertainties, the key to adaptation must be resilience – managing risks and building capacity to deal with unpredictable events, especially of the most vulnerable rural and urban poor. To build resilience to ongoing and future climate change calls for immediate adaptation.

In WWDR 2012, which was wholly dedicated to the theme 'Managing water under uncertainty and risk', it says:

> Risk and uncertainty characterize much of what water managers and socio-economic policy-makers must deal with. The more they understand these uncertainties and risks, the more effectively they can plan,

> design and manage water systems to reduce these risks and uncertainties.
>
> (WWDR, 2012, p. 11)

However, there is also a sentiment that uncertainties and risks cannot be planned, designed, or managed away, but that efforts should be made to reduce the impacts on unforeseen development. A resilient approach in water governance has therefore been discussed in terms of a 'safe to fail' ethos (Butler *et al.*, 2016, building on Holling, 1996). Such an ethos acknowledges that systems *will* fail but, instead of focusing on the threats that cause such failures, emphasis is placed on managing and overcoming those failures (ibid.).

The importance of a resilient approach is recognized by WHO. According to that organization, unless resilience is achieved in water supply we will face 'serious public health consequences as water quality deteriorates, water quantity becomes less certain and sanitation systems cause environmental contamination' (WHO, 2009, p. 19). The water and resilience approaches and literature are not univocal, however. In effect, there are contestations over the way the concept is understood and operationalized and different resilience frameworks exist in the literature (Butler *et al.*, 2016; see also Earle *et al.*, 2015, for the climate change debate in relation to transboundary water governance).

However, a number of policy responses have been suggested by WHO. The organization argues that challenges related to climate change make it even more important to integrate issues of drinking water and sanitation into IWRM policies and practices, all the more so since IWRM offers an approach which can aid the process of assessing and negotiating water needs and access among competing demands and uses (WHO, 2009). In terms of the institutional framework, centralized management approaches are argued to have a positive impact on resilience since high technical competence and economies of scale in administration will contribute to adaptive capacity, in comparison to systems which are very decentralized. Such systems are, in turn, related to 'high rates of failure' (ibid., p. 25).

This means that household piped water from utility-managed systems is seen as desirable for creating possibilities for scaling-up resilience. However, such utility-based supplies can also be vulnerable because of their size and complexity. A number of factors determine whether they are resilient or not, including how well-trained staff are and whether there is financial capital available for investing in technology and upgrading infrastructure (ibid., 2009, p. 16). In working towards resilience, it is also important: to reduce leakage; to spread risk by using multiple and independent sources; to use reservoirs and aquifers to store water in order to safeguard water accessibility in times of inter-seasonal variability; to implement sustainable management of groundwater sources; and to work with preventive management approaches (ibid., 2009, p. 25). Another suggestion for creating resilience of water accessibility is to increase supply through desalination of coastal water/brackish water. Here, however, it is noted that more research is needed to find ways of providing such systems with low carbon energy. Even if piped household water is preferred to promote health and easy access as well as resilience, community-managed water is a reality for a big share of the world's population. Among the community-managed sources, tube wells are those that are considered most resilient in comparison to dug wells or springs (ibid.).

The question is: what is at stake in this shift to resilience in water governance? As was discussed in Chapter 2, it involves a shift in the understanding of the ontologies of both the environment and human subjects. It has been suggested that resilience is a concept that works in tandem with neoliberal techniques of governing that aim to create self-governing and self-reliant subjects who are ready for change (Evans and Reid, 2014; Reid, 2012, 2013; see also Hellberg and Knutsson, 2016). As has been shown above, it is evident that the concept of resilience has found a home in the water context. Because of the essential need for water, water governance and water management always have to manage risks and threats in connection to both quantity and quality of water. Water and water access are also particularly vulnerable to risks associated with climate change, such as flood and drought.

Yet, what seems to be argued for, at least in the influential publication by WHO (2009), is a resilience approach that aims at *including* populations in utility-based services rather than leaving them to live 'within the limits of their own self-reliance' (Duffield, 2007, p. 68). It seems, thus, that the concept of resilience, in this context, can be used to push for scaling-up efforts for water services to those who lack services, and especially to those who rely on community-based sources, which are deemed to be the least resilient. In other words, a resilience approach, as described above, is noteworthy since such efforts aim to create a situation for the poor that is the opposite of self-reliance.

However, it might be the very *biopolitics* of water that can explain such a seemingly counter-intuitive conclusion. Water is exceptional because it is absolutely essential for survival. Here, Duffield's and Foucault's stipulations help us grasp the biopolitics at play. It is central to the biopolitics of water to differentiate between those who are expected to be content with water for basic needs and those who can enjoy a redundant access to it as long as they can pay. Thus, in the water context, it is above the threshold of survival that people are left to their own self-reliance, to be their own 'enterprise' (Foucault, 2008, p. 206).

Conclusion

This chapter has outlined the contours of GWG. It has described how a new water management discourse emerged in the 1990s as a response to new ways of understanding water challenges. In this new story around water, global water resources were seen as being in the midst of a crisis, due to population growth, urbanization, industrialization, and an uncertain supply related to the pollution of water resources and climate change. In this narrative, water scarcity was depicted as a central problem. That many people were still lacking adequate access to safe water and that unsustainable uses of water had a degrading effect on ecosystems, were also seen as important water challenges.

Integrated, participatory, gender-sensitive, and demand-side management, which was supposed to be both economically and ecologically sound, increasingly became the answer to these challenges. The framework is known as Integrated Water Resources Management (IWRM). While the IWRM framework has dominated the discussion of water governance for the last decades, contemporary GWG has embraced the concept of water security as an important component. Although there are major differences between the frameworks, the water security concept, at least in a broad understanding, overlaps with both the definition and the focus areas of IWRM. One of these areas is the emphasis that is placed on water's importance in guaranteeing basic human needs, something which was also reflected in the MDGs and is now a component of the 2030 agenda. In this chapter, meeting basic human needs has been identified as part of the main rationale for contemporary GWG, together with the importance of guaranteeing water for economic development and for ecosystems.

A biopolitical reading has explored how different populations are viewed, within the frames of the three main governing rationales, and the role that scarcity plays in this governance problematique. Even though GWG should in no way be seen as consistent, but rather as complex and sometimes even contradictory in the way that it frames water governance issues, the chapter has outlined a few central biopolitical aspects of such governance. One important conclusion is that poor people's relationship to water management, sustainability, and security has clear linkages with Malthus' stipulations. This is because of how the disequilibrium between 'the poorest' populations and 'ecosystem services' is assumed to result in both food and water insecurities for these poor populations and simultaneously create a risk of (violent) conflict.

This preoccupation with describing the poor as confined to a particular territory with little possibility to turn the present availability of resources into future sustainability keeps reoccurring in sustainable development discourse. It can be explained by the 'Malthus effect' (cf. Dean, 2015) as it is inscribed in liberal economic discourse. Thus, Malthus lives on and produces meaning around the 'ecologically ignorant poor' (cf. Reid, 2013) in GWG. In turn, this reflects global hierarchizations of how different forms of (un)sustainable and (ir)responsible forms of lives are understood.

The chapter has, furthermore, outlined how the focus on meeting basic human needs through a human right to water can be problematized, because of how such an agenda involves a biopolitical separation between different populations, that is, between those who are expected to rely merely on water for basic needs and those who can access larger amounts. Moreover, the way that the ecological dimension of water management is understood in the GWG reports elucidates how present sustainable development discourse in general, and GWG in particular, are based on a biopolitical distinction between humans and the rest of nature. In fact, through the way that ecosystems, and everything that is contained in such a vast concept, are described as a resource base for economic development and human needs, the idea of 'nature', or something other than human, almost disappears altogether. Thus, GWG is characteristically biopolitical in how it aims to govern all of life and how it constructs the population–resource problematique and the idea of futurity in relation to different populations, as well as how it differentiates between different forms of human and non-human lives.

Moreover, we can understand GWG as *neoliberal* in a governmentality sense. It is *regulatory*, which means that it is a form of governance that is directed at governing *processes* and *people* at a general level. It is, furthermore, a form of governance that takes the principles of *market economy* as the model for the proper allocation of water (this despite the fact that it is recognized, in several instances in WWDR reports, that growth and economic development pose challenges to sustainability, both ecologically and socially).

However, labelling GWG as neoliberal does not mean that such governance argues that water should be put on the market as a commodity like any other. A governmentality understanding posits the importance of interventions in market processes for competitive mechanisms to be able to play a regulatory role (Foucault, 2008). The inclusion of ecosystem services in conventional economic models and the adoption of the (human) right to water can be understood as such interventions. What is meant is that, in GWG, water is to be governed for its most economically beneficial use, with the exception that governance interventions are to guarantee, at a minimum level, water for the poor and for the environment. It should also be noted that the actual governance outcomes in different local, national, and regional settings are to be treated as an empirical question and in relation to what particular governance models are adopted in each context. The next chapter turns to this, and to particular 'hydromentalities'.

Notes

1 For a more exhaustive list of global water organizations see Cooley *et al.* (2013, p. 14).

2 It has, however, been argued that these figures underestimate water scarcity because they do not capture seasonal fluctuations in consumption and availability. When taking such fluctuations into account, Mekonnen and Hoekstra argue that as many as four billion people experience water scarcity for at least one month per year (Mekonnen and Hoekstra, 2016; WWDR, 2017).

3 There are no figures on institutional water scarcity and no updated figures on economic water scarcity.

References

2030 WRG (Water Resources Group). (2009) *Charting our Water Future: Economic frameworks to inform decision-making*. Washington, DC.

Ahlers, R. and Zwarteveen, M. (2009) 'The water question in feminism: water control and gender inequalities in a neoliberal-era', *Gender, Place and Culture: A Journal of Feminist Geography*, 16(4), 409–426.

Anokye, N. A. and Gupta, J. (2012) 'Reconciling IWRM and water delivery in Ghana – The potential and the challenges', *Physics and Chemistry of the Earth*, 47–48, 33–45.

Arendt, H. (1998) *The Human Condition* (2nd edn). Chicago, IL: University of Chicago Press.

Bakker, K. (2010) *Privatizing Water: Governance Failure and the World's Urban Water crisis*. Ithaca, NY: Cornell University Press.

Benjaminsen, T. A. and Ba, B. (2009) 'Farmer-herder Conflicts, Pastoral Marginalisation, and Corruption: A Case Study from the Inland Niger Delta of Mali', *Geographical Journal*, 175(1), 71–81.

Benson, D., Gain, A. K., and Rouillard, J. J. (2015) 'Water governance in a comparative perspective: From IWRM to a 'nexus' approach?', *Water Alternatives*, 8(1), 756–773.

Birkenholtz, T. (2013) ' "On the network, off the map": developing intervillage and intragender differentiation in rural water supply', *Environment and Planning D: Society and Space*, 31(2), 354–371.

Biswas, A. K. (2004) 'Integrated Water Resources Management: A reassessment', *Water International*, 29(2), 248–256.

Boelee, E. (ed.) (2011) *Ecosystems for Water and Food Security*. Nairobi/Colombo, United Nations Environment Programme (UNEP)/International Water Management Institute (IWMI).

Bohensky, E. L. (2008) 'Discovering resilient pathways for South African water management: two frameworks for a vision', *Ecology and Society*, 13(1), 19.

Butler, D., Farmani, R., Fu, G., Ward, S., Diao, K., and Astaraie-Imani, M. (2014) 'A new approach to urban water management: safe and sure', *Procedia Engineering*, 89, 347–354.

Buur, L., Jensen, S. and Stepputat, F. (2007) *The Security–Development Nexus: Expressions of Sovereignty and Securitization in Southern Africa*. Uppsala, Sweden: The Nordic Africa Institute.

CBD (Convention on Biological Diversity). (2014) Ecosystem Approach, www.cbd.int/ecosystem/

Chikozho, C. (2008) 'Globalizing integrated water resources management: a complicated option in Southern Africa', *Water Resources Management*, 22(9), 1241–1257.

Cleaver, F. and Hamada, K. (2010) ' "Good" water governance and gender equity: a troubled relationship', *Gender and Development*, 18(1).

Cook, B. R and Spray, C. J. (2012) 'Ecosystem services and integrated water resource management: Different paths to the same end?' *Journal of Environmental Management*, 109, 93–100

Cook, C. and Bakker, K. (2012) 'Water security: Debating an emerging paradigm', *Global Environmental Change*, 22(1), 94–102.

Cooley, H., Ajami, N., Ha, M.-L., Srinivasan, V., Morrison, J., Donnelly, K., and Christian-Smith, J. (2013) *Global Water Governance in the 21st century*. Oakland, CA: Pacific Institute.

Costanza, R., de Groot, R., Sutton, P., van der Ploeg, S., Anderson, S. J., Kubiszewski, I., Farber, S., and Turner, R. K. (2014) 'Changes in the global value of ecosystem services', *Global Environmental Change*, 26, 152–158.

Crow-Miller, B., Webber, M., and Molle, F. (2017) 'The (re)turn to infrastructure for water management?', *Water Alternatives*, 10(2), 195–207.

Dean, M. (2015) 'The Malthus Effect: population and the liberal government of life', *Economy and Society*, 44(1), 18–39.

DFID (Department for International Development). (2005) *Fighting Poverty to Build a Safer World: A Strategy for Security and Development*. London, UK: DFID.

Duffield, M. (2007) *Development, Security and Unending War: Governing the World of Peoples*. Cambridge, UK: Polity.

Earle, A., Cascão, A. E., Hansson, S., Jägerskog, A., Swain, A., and Öjendal, J. (2015) *Transboundary Water Management and the Climate Change Debate*. London, UK: Routledge.

Evans, B. and Reid, J. (2014) *Resilient Life: The art of living dangerously*. Cambridge, UK: Polity.

FAO (Food and Agriculture Organization). (2012) *Coping with Water Scarcity: An Action Framework for Agriculture and Food Security*. FAO Water Reports No. 38. Rome: FAO. www.fao.org/docrep/016/i3015e/i3015e.pdf

FAO. (2016) *Coping with water scarcity in agriculture: a global framework for action in a changing climate*, www.fao.org/37a-i5604e.pdf [accessed 28 September 2017].

Foucault, M. (2008) *The Birth of Biopolitics, Lectures at the Collège de France 1978–1979*. New York, NY: Palgrave Macmillan.

Giordano, M. F., Giordano, M. A., and Wolf, A. T. (2005) 'International Resource Conflict and Mitigation', *Journal of Peace Research*, 42(1), 47–65.

Gleick, P. H. (ed.) (1993) *Water in Crisis: A Guide to the World's Freshwater Resources*. New York, NY: Oxford University Press.

Grey, D. and Sadoff, C. W. (2007) 'Sink or Swim? Water security for growth and development', *Water Policy*, 9, 545–571.

Gudmand-Høyer, M. and Lopdrup Hjorth, T. (2009) 'Liberal Biopolitics Reborn. Review Essay', *Foucault Studies*, 7, 99–130.

GWI (Global Water Institute). (2013) *Future Water (In)security: Fact, Figures and Predictions*, www.gwiwater.org/sites/default/files/pub/FUTURE%20WATER%20(IN)SECURITY.pdf [accessed 28 September 2017].

GWP (Global Water Partnership). (2000a) *TAC Background Paper No. 4: Integrated Water Resources Management*. Denmark: Global Water Partnership.

GWP. (2000b) *Towards Water Security: A Framework for Action Global Water Partnership*. Stockholm, Sweden.

GWP. (2017) *IWRM Toolbox*, www.gwp.org/en/learn/iwrm-toolbox/About_IWRM_ToolBox/ [accessed 28 September 2017].

Hall, R. P., van Koppen, B., and van Houwling, E. (2014) 'The human right to water: The importance of domestic and productive water rights', *Science and Engineering Ethics*, 20(4), 849–868.

Harris, L. (2008) 'Water Rich, Resource Poor: Intersections of Gender, Poverty, and Vulnerability in Newly Irrigated Areas of Southeastern Turkey', *World Development*, 36(12), 2643–2662.

Harris, L. (2009) 'Gender and emergent water governance: comparative overview of neoliberalized natures and gender dimensions of privatization, devolution and marketization', *Gender, Place and Culture*, 16(4), 387–408.

Hellberg, S. (2014) 'Water, life and politics: Exploring the contested case of eThekwini municipality through a governmentality lens', *Geoforum*, 56, 226–326.

Hellberg, S. (2017) 'Water for survival, water for pleasure – A biopolitical perspective on the social sustainability of the basic water agenda', *Water Alternatives*, 10(1), 65–80.

Hellberg, S. and Knutsson, B. (2016) 'Sustaining the life-chance divide? Education for sustainable development and the global biopolitical regime', *Critical Studies in Education*. DOI: http://dx.doi.org/10.1080/17508487.2016.1176064

Holling, C. S (1996) 'Engineering Resilience versus Ecological Resilience', in P. Schulze (ed.), *Engineering Within Ecological Constraints*. National Academy of Engineering.

Homer-Dixon, T. (1994) 'Environmental Scarcities and Violent Conflict: Evidence from Cases', *International Security*, 19(1), 5–40.

Homer-Dixon, T. (1999) *Environment, Scarcity, and Violence*. Princeton, NJ: Princeton University Press.

ICWE (International Conference on Water and the Environment). (1992) *The Dublin Statement on Water and Sustainable Development*. Dublin, IE: ICWE.

Jeffrey, P. and Geary, M. (2006) 'Integrated water resources management: lost on the road from ambition to realisation?', *Water Science and Technology*, 53(1), 1–8.

Kaplan, R. D. (1994) 'The Coming Anarchy', *Atlantic Monthly*, 273, 44–77.

Kujinga, K. (2002) 'Decentralizing water management: An analysis of stakeholder participation in the management of water in the Odzi subcatchment area, Save Catchment', *Physics and Chemistry of the Earth*, 27, 897–905.

Lenton, R. and Muller, M. (2009) *Integrated Water Resources Management in Practice: Better water management for development*. London, UK: Earthscan.

Linton, J. (2010) *What is Water? The History of a Modern Abstraction*. Vancouver, BC: UBS Press.

Loftus, A. (2015) 'Water (in)security: securing the right to water', *The Geographical Journal*, 181(4), 350–356.

McDonnell, R. A. (2008) 'Challenges for integrated water resources management: how do we provide the knowledge to support truly integrated thinking?', *International Journal of Water Resources Development*, 24, 131–143.

MEA (Millennium Ecosystem Assessment). (2005) *Ecosystems and Human Well-Being: Wetlands and Water Synthesis*. Washington, DC: World Resources Institute (WRI).

Mekonnen, M. M. and Hoekstra, A. Y. (2016) 'Four billion people facing severe water scarcity', *Science Advances*, 2(2).

Mildner, S.-A., Lauster, G. and Wodni, W. (2011) 'Scarcity and Abundance Revisited', *International Journal of Conflict and Violence*, 5(1), 155–172.

Mitchell, B. (2006) 'IWRM in practice: lessons from Canadian experiences', *Journal of Contemporary Water Research and Education*, 51–55.

Molle, F. (2008) 'Nirvana concepts, narratives and policy models: insights from the water sector', *Water Alternatives*, 1(1), 23–40.

Molle, F. (2009) 'Water and society: new problems faced, new skills needed', *Irrigation and Drainage* 58, 205–211.

Moseley, W. (1991) 'African evidence on the relation of poverty, time preference and the environment', *Ecological Economics*, 38(3), 317–326.

Mostert, E. (2003) 'The Challenge of Public Participation', *Water Policy*, 5, 179–197.

Moyo, S. (2005) 'Land and Natural Resource Redistribution in Zimbabwe: Access, Equity and Conflict', *African and Asian Studies*, 4(1–2), 187–223.

Njoh, A. J. (2002) 'Barriers to community participation in development planning: lessons from the Mutengene (Cameroon) self-help water project', *Community Development Journal*, 37(3), 233–248.

O'Reilly, K. (2006) ' "Traditional" women, "Modern" water: Linking gender and commodification in Rajasthan, India', *Geoforum*, 37(6), 958–972.

OHCHR (Office of the United Nations High Commissioner for Human Rights). (2010) 'The Right to Water', Fact Sheet No. 35. Geneva, Switzerland: United Nations.

Öjendal, J., Hansson, S., and Hellberg, S. (2012) *Politics and Development in a Transboundary Watershed: The Case of the Lower Mekong Basin*. Dordrecht, NL: Springer.

Ozerol, G. and Newig, J. (2008) 'Evaluating the success of public participation in water resources management: five key constituents', *Water Policy*, 10, 639–655.

Pahl-Wostl, C., Gupta, J., and Petry, D. (2008) 'Governance and the Global Water System: Towards a Theoretical Exploration', *Global Governance*, 14, 419–436.

Petit, O. and Baron, C. (2009) 'Integrated water resources management: From general principles to its implementation by the state. The case of Burkina Faso', *Natural Resources Forum*, 33, 49–59.

Prokopy, L. S. (2005) 'The relationship between participation and project outcomes: Evidence from rural water supply projects in India', *World Development*, 33(11), 1801–1819.

Rahaman, M. and Varis, O. (2005) 'IWRM: evolution, prospects and future challenges', *Sustainability: Science, Practice, and Policy*, 1(1), 15–21.

Reid, J. (2012) 'The disastrous and politically debased subject of resilience', *Development Dialogue*, 58, April 2012.

Reid, J. (2013) 'Interrogating the Neoliberal Biopolitics of the Sustainable Development-Resilience Nexus', *International Political Sociology*, 7(4), 353–367.

Rockström, J., Falkenmark, M., Folke, C., Lannerstad, M., Barron, J., Enfors, E., Gordon, L., Heinke, J., Hoff, H., and Pahl-Wostl, C. (2014) *Water Resilience for Human Prosperity*. Cambridge, UK: Cambridge University Press.

RSA (Republic of South Africa). (1975) *Report on the Crocodile River (Eastern Transvaal) Government Waterwork (Sterkspruit Dam)*, Doc. No. W.P. T-'75. Pretoria: Government Printer.

RSA (Republic of South Africa). (1996) Constitution of the Republic of South Africa, Act 108 of 1996. Pretoria: Government Printer.

Saad-Filho, A. and Weeks, J. (2013) 'Curses, Diseases and Other Resource Confusions', *Third World Quarterly*, 34(1), 1–21.

Savenije, H. H. G. and Van der Zaag, P. (2008) 'Integrated water resources management: concepts and issues', *Physics and Chemistry of the Earth*, 33, 290–297.

Srinivasan, V., Konar, M., and Sivapalan, M. (2017) 'A dynamic framework for water security', *Water Security*. DOI: http://dx.doi.org/10.1016/j.wasec.2017.03.001.

Starr, J. (1991) 'Water Wars', *Foreign Policy*, 82, 17–36.
Stern, M. and Öjendal, J. (2010) 'Mapping the Security–Development Nexus: Conflict, Complexity, Cacophony, Convergence?', *Security Dialogue*, 41(1), 5–30.
Sultana, F. (2009) 'Fluid lives: subjectivities, gender and water in rural Bangladesh', *Gender, Place and Culture*, 16(4), 427–444.
Sultana, F. (2011) 'Suffering for Water, Suffering from Water: Emotional Geographies of Resource Access, Control and Conflict', *Geoforum*, 42(2), 163–172.
Swyngedouw, E. (2004) *Social Power and the Urbanization of Water: Flows of Power*. Oxford, UK: Oxford University Press.
Swyngedouw, E. (2013) 'UN Water Report 2012: Depoliticizing Water', *Development and Change*, 44(3), 823–835.
Swyngedouw, E. (2015) *Liquid Power: Contested Hydro-Modernities in Twentieth-Century Spain*. Cambridge, Massachusetts/London, England: MIT Press.
Tellmann, U. (2013) 'Catastrophic Populations and the Fear of the Future: Malthus and the Genealogy of Liberal Economy', *Theory, Culture and Society*, 30(2), 135–155.
Theisen, O. M. (2008) 'Blood and Soil? Resource Scarcity and Internal Armed Conflict Revisited', *Journal of Peace Research*, 45(6), 801–18.
UN (United Nations). (2004) *A More Secure World: Our Shared Responsibility*. Report of the Secretary-General's High-Level Panel on Threats, Challenges and Change. New York, NY: United Nations.
UN (United Nations). (2008) *Official list of MDG indicators*, https://mdgs.un.org/unsd/mdg/Resources/Attach/Indicators/OfficialList2008.pdf [accessed 22 September 2017].
UN (United Nations). (2010) Resolution 64/292. *The human right to water and sanitation*, www.un.org/en/ga/search/view_doc.asp?symbol=A/RES/64/292 [accessed 27 September].
UN (United Nations). (2017a) *Millennium development goals and beyond 2015*, www.un.org/millenniumgoals/ [accessed 27 September 2017].
UN (United Nations). (2017b) *Sustainable Development Knowledge Platform*, https://sustainabledevelopment.un.org/?menu=1300 [accessed 27 September 2017].
UN SDG 6. (2017) https://sustainabledevelopment.un.org/sdg6 [accessed 26 September 2017].
UN Water. (2007) *Coping with water scarcity: Challenge of the twenty-first century*. World Water Day, 22 March 2007.
UN Water. (2010) *Climate Change Adaptation: The Pivotal Role of Water*. Policy and analytical brief, www.unwater.org/publications/climate-change-adaptation-pivotal-role-water/ [accessed 6 October 2017].
UN Water. (2014) *Delivering as One on Water Related Issues. Strategy 2014–2020*, www.unwater.org/publications/un-water-strategy-2014-2020/ [accessed 19 September 2017].
UN Water. (2017a) *What we do*, www.unwater.org/what-we-do/inform-policies/ [accessed 28 September 2017].
UN Water. (2017b) *Scarcity*, www.unwater.org/water-facts/scarcity/ [accessed 28 September 2017].
UN Water. (2017c) *Human Rights*, www.unwater.org/water-facts/human-rights/ [accessed 28 September 2017].
UN Water. (2017d) *Ecosystems*, www.unwater.org/water-facts/ecosystems/ [accessed 28 September 2017].

UNSD (United Nations Statistics Division). (2016) *Metadata Goal 6*, https://unstats.un.org/sdgs/files/metadata-compilation/Metadata-Goal-6.pdf [accessed 21 September 2017].

UNSD. Millennium Development Goals Indicators Database. The official United Nations site for the MDG Indicators, http://unstats.un.org/unsd/mdg/Metadata.aspx?IndicatorId=0&SeriesId=768 [accessed 28 September 2017].

UNU (United Nations University). (2013) *Water Security and the Global Water Agenda*. A UN Water Analytical Brief, www.unwater.org/publications/water-security-global-water-agenda/ [accessed 26 September 2017].

van Koppen, B., Smits, S., Moriarty, P., Penning de Vries, F., Mikhail, M. and Boelee, E. (2009) *Climbing the Water Ladder: Multiple-use water services for poverty reduction*. TP Series No. 52. The Hague, the Netherlands: IRC International Water and Sanitation Centre and International Water Management Institute, www.iwmi.cgiar.org/Publications/Other/PDF/TP52_Climbing_2009.pdf? Galog =no [accessed 15 January 2017].

Varis, O., Rahaman, M. and Stucki, V. (2008) 'The rocky road from integrated plans to implementation: lessons learned from the Mekong and Senegal river basins', *International Journal of Water Resources Development*, 24(1), 103–121.

WCED (World Commission on Environment and Development). (1987) *Our Common Future*. Oxford: Oxford University Press.

WGF (Water Governance Facility). (2012) *Human Rights-Based Approaches and Managing Water Resources: Exploring the potential for enhancing development outcomes*. WGF Report No. 1. Stockholm, SE: Stockholm International Water Institute (SIWI).

WHO (World Health Organization) and UNICEF (United Nations Children's Fund). (2015) Joint Monitoring Program (JMP): *Progress on sanitation and drinking water: 2015 Update and MDG assessment*, www.wssinfo.org/fileadmin/user_upload/resources/JMP-Update-report-2015_English.pdf [accessed 31 August 2016].

WHO (World Health Organization). (2009) *Vision 2030: The resilience of water supply and sanitation in the face of climate change*. Geneva, Switzerland: WHO.

Wolf, A. T., Yoffe, S. B., and Giordano, M. (2003) 'International waters: Identifying basins at risk', *Water Policy*, 5, 29–60.

Wolfe, C. (2012) *Before the Law: Humans and Other Animals in a Biopolitical Frame*. Chicago, IL: University of Chicago Press.

World Bank. (2004) *Water Resources Sector Strategy: Strategic directions for World Bank engagement*. Report No 28114, http://documents.worldbank.org/curated/en/941051468765560268/pdf/281140PAPER0Water0Resources.pdf [accessed 2 October 2017].

WWDR (World Water Development Report). (2003) *Water for People, Water for Life*. UNESCO, http://unesdoc.unesco.org/images/0012/001297/129726e.pdf [accessed 8 October 2017].

WWDR (World Water Development Report). (2006) *Water – a shared responsibility*. UNESCO, http://unesdoc.unesco.org/images/0014/001454/145405E.pdf [accessed 8 October 2017].

WWDR (World Water Development Report). (2009) *Water in a Changing World*. UNESCO, http://unesdoc.unesco.org/images/0018/001819/181993e.pdf [accessed 8 October 2017].

WWDR (World Water Development Report). (2012) *Managing Water under Uncertainty and Risks*. UNESCO, www.unesco.org/new/en/natural-sciences/environment/water/wwap/wwdr/wwdr4-2012/ [accessed 8 October 2017].

WWDR (World Water Development Report). (2015) *Water for a Sustainable World*. UNESCO, http://unesdoc.unesco.org/images/0023/002318/231823E.pdf [accessed 8 October 2017].

WWDR (World Water Development Report). (2016) *Water and Jobs*. UNESCO, http://unesdoc.unesco.org/images/0024/002439/243938e.pdf [accessed 8 October 2017].

WWDR (World Water Development Report). (2017) *Wastewater the Untapped Resource*. UNESCO, http://unesdoc.unesco.org/images/0024/002471/247153e.pdf [accessed 8 October 2017].

WWF (World Wildlife Fund) (2012) *Living Planet Report 2012: Biodiversity, Biocapacity and Better Choices*. Gland, Switzerland: WWF international.

4 Hydromentalities

Introduction

While Chapter 3 analysed the biopolitical rationalities of present global water governance as they appear in policy, this chapter explores water governance from a broader perspective, addressing not only different ways of thinking about and defining water, but also the techniques involved in its governance as they take shape in different historical and geographical contexts. Thereby this study of water governance continues from a (green) governmentality perspective, which emphasises how governing is part of producing 'reality' in terms of constituting agents, their interests and identities, as well as through employing and legitimizing certain forms of rule.

This chapter employs the concept of 'hydromentality', which is defined in this book as assemblages of governing rationalities, techniques of rule and ways of thinking about, and defining, water (partly building on Mayhew, 2004 and Linton, 2010). Hydromentalities involve: particular ways of understanding what water is; the challenges involved in its allocation; the proper way to manage and govern the water, water use, and water users.

By this means the chapter offers a perspective on how we can understand present global water governance as a neoliberal hydromentality. However, the chapter also makes the case that in order to understand and assess the impact of (global) water governance, we need to conduct careful empirical investigations of the ways in which general and global guidelines are translated into national and local policies and practices, and their effects.

The first section outlines the characteristics of two major paradigms of water governance: the hydraulic mission and a neoliberal hydromentality. The second section provides an overview of governmentality studies of water governance in various contexts, from Ecuador, India, Paraguay, Niger, and China, and from Jakarta, Indonesia. The third section outlines the particular hydromentality of eThekwini municipality, South Africa.

Hydromentalities: major paradigms

This first section discusses two paradigms of water and the shift between them, but two things need to be made clear beforehand. First, the narrative that is outlined here starts with a discussion of the, so-called, state hydraulic paradigm/the hydraulic mission. Such a hydromentality connects the role of water to modernity and the construction of the nation state and its citizens, and it is both gendered and related to issues of class. As will become clear from the South African example, it is also colonial, both in relation to people who are not seen as part of the state's civilizing and developmental mission and in relation to non-human lives. This way of opening the story about hydromentalities should not be understood as asserting that there are, or have been, no other hydromentalities. Rather, the point is to provide a perspective on the trajectory of the rationalities, mentalities, and techniques of present-day dominant policy prescriptions. Overall, this chapter argues that dominant governance discourse 'others' hydromentalities different from its own (see also Linton, 2010, p. 14), at the same time as wanting to incorporate a multiplicity of viewpoints, especially in terms of the value of water. Second, and relatedly, while this first section describes a shift in water governance paradigms as one that has taken place in time, we should be wary of how logics transcend the different paradigms. In other words, there is not a clear-cut line between the different paradigms. As mentioned in the previous chapter, a renewed focus on water supply through waste water reuse and desalination can be viewed as either a *continuance* of rationalities from the hydraulic mission or as a *(re)turn* to big (and costly) infrastructural solutions (Crow-Miller *et al.*, 2017).

The hydraulic mission

Water regulation during the twentieth century can, in many parts of the world, be characterized as belonging to the, so-called, state hydraulic paradigm, or the 'municipal hydraulic paradigm' (Bakker, 2005, 2010), also called the 'hydraulic mission' (Molle *et al.*, 2009) (which is the term used henceforth in this book).

This paradigm was characterized by a belief that there is an abundance of water that can meet the growing needs of a modernizing society through hydraulic technologies (Bakker, 2010). In this paradigm, state/public ownership of water resources and water services provision was the norm and was justified by the large-scale investments that were needed to construct the infrastructure for the abstraction of water and for providing access. In this paradigm, water was considered essential to a nation's development and to a dignified living standard for its citizens (ibid., pp. 34–35).

In Jamie Linton's (2010) terminology, the understanding of water in this paradigm belongs to 'modern water'. The political work in relation to the representation of water within this paradigm is based on the assumption that 'any and all waters can be and should be considered apart from their social and ecological relations and reduced to an abstract quantity' (ibid., p. 14). Hence, in this discourse:

> all water is made known as an abstract, isomorphic, measurable quantity that may be reduced to its fundamental unit – a molecule of H_2O – and represented as the substance that flows in the hydrological cycle.
>
> (Ibid.)

Through this scientific understanding, water became rendered as a 'natural fact' (ibid., p. 184), a resource that could be made available for human use. Borrowing from Bruno Latour, Linton argues that the double contradiction of upholding the separateness of society and water while simultaneously being dependent on water for 'virtually every aspect of social production' has the effect of making 'it possible to do everything [with water] without being limited by anything' (ibid., citing Latour, 1993, p. 32). This is possible since water resources, water scarcities, and water purity are understood as 'perfectly natural' (Linton, 2010, p. 184). Relatedly, Molle *et al.* (2009) – who term this paradigm 'the hydraulic mission' – state:

> In other words, all the ingredients of the 19th century scientism, hubris and utopias were alive and well. Beyond the objectives of food/energy production or self-sufficiency the hydraulic mission at its height was a celebration of technology and dominations over nature, a linear view of history based on Rostowian stages – where mechanisation intensifications and economics of scale would replace backward, if not 'barbarous' practices
>
> (Molle *et al.*, 2009, p. 336)

The hydraulic mission was thus closely related to modernization and nation building as well as having a civilizing mission.

South Africa is an illustrative example of how the hydraulic mission took shape and was strongly connected to the creation of a racist state. Colonization had destroyed communal ownership of land and natural resources such as water and had introduced a system of private ownership. When the British took over from the Dutch in 1805, a permanent tenure system was formalized, as was the so-called riparian principle, which granted control of water to the landowner (Funke *et al.*, 2007). As the settlers expanded to the more arid interior of present-day South Africa, the question of water became more urgent. Therefore, emphasis was placed on irrigation to support agriculture. Also, the discovery of gold and diamonds placed a higher demand on water resources in mining areas, and mining operations were subsequently granted water rights with priority over other uses (Funke *et al.*, 2007, p. 13; Muller, n.d., pp. 12, 14).

Given the centrality of the water question for the European settlers, experts were appointed to study water availability and to give recommendations about how to address water-related problems (Muller, n.d., p. 11). Large-scale control of management of water resources was already in the minds of water experts at the end of the nineteenth century. Such an early devotion to a hydraulic mission reflects the approach that South Africa took during the twentieth century (Muller, n.d.).

The main stimuli of the hydraulic mission for twentieth-century South Africa were the drought and depression of the 1930s that had a severe effect on the country's agriculture (Turton *et al.*, 2004, p. 399). Not only were water schemes a way of satisfying growing needs for water, but these water resource development projects were also a means of providing jobs for white workers, which functioned so as to alleviate 'the poor white problem that had been haunting South Africa since the 1890s' (ibid.).[1]

During this whole period, the black population had been neglected, and their situation worsened as the Nationalist Party (NP) came to power in 1948 (Funke *et al.*, 2007, p. 13). Although some investments were made in irrigation and farmer development in the homelands (Muller, n.d.), the black majority did not experience the benefits of developments in the water sector. As a result of apartheid, its dispossession and forced removals, existing farming activities were distorted (ibid.) and instead of being granted basic rights, the homelands had to negotiate water rights in competition with users outside their territories (Funke *et al.*, 2007, p. 13, drawing on RSA, 1975). During this time, the NP government used 'instruments of state ownership and policy to expand the economy and make it more self-sufficient' (Muller, n.d., p. 12). In terms of water, this meant that a previous focus on irrigation shifted to a focus on industry and the development of urban areas (ibid.). Thus, water was central to nation building in South Africa during the apartheid years and was also 'a very effective weapon in the apartheid government's arsenal of oppression and control' (Funke *et al.*, 2007, p. 13).

At the same time as increasing the number of large water projects for socio-economic development (which benefited the white minority), the state gradually increased its control over water resources during the 1930s and 1940s. This development culminated in the 1956 Water Act. The Act established the control, allocation, and use of water in public streams, which 'enabled the country to engage in a major programme of water resources development' (Muller, n.d., p. 15). This commitment can be illustrated by the words of P. K. Le Roux, Minister of Water Affairs, in March 1962, who outlined the plans for the construction of the Orange River project:[2]

> The Government has therefore decided to undertake the biggest, most important and most spectacular water supply project ever initiated in the history of our country's water affairs. Not only is it the biggest in

> Africa, when seen as a whole, but it will be one of the biggest projects of its kind in the world.

Furthermore, Le Roux declared that:

> In the history of all young civilised countries the time arrives when big and imaginative water development projects must be launched to promote the growth of areas of development, the formation of industries and the generation of electric power, and to create a means of coping with the future population increase, so as to maintain the rate of progress for the country as a whole. That is the principal aim of the Orange River Project.
> (RSA, 1962, p. 3, quoted in Turton *et al.*, 2004, pp. 183–184)

That South Africa was to follow the global trend of massive water projects was clear in Le Roux's speech, and he linked the project to other major dam projects around the world, such as the Boulder Dam in the USA and the Kariba Dam on the Zambezi. He continued:

> The Government feels that the time has now come for the Republic of South Africa also to initiate a giant water supply project in order to create conditions in which development can take a big stride forward, and timeously and systematically provide for the country's future requirements in the sphere of agricultural, industrial and urban [water] use.
> (Ibid., p. 184)

In these quotations, big water supply infrastructure is connected to notions of civilization, progress, and development, and used to represent the power and capacity of the apartheid state.

In South Africa, as elsewhere, dams and reservoirs became the ultimate symbols of modernization and nation building during the hydraulic mission (Bakker, 2012, p. 618). In turn, their 'domestic counterparts' were 'the washbasin and the private flush toilet in the boudoir' (ibid.). Critical geographers such as Maria Kaika (2005) and Matthew Gandy (2004, 2006, 2014) have described how water has played a central role in the way that relationships between modernity, nature, and urban life have unfolded. While water in early-nineteenth-century Western cities, such as London, presented a threat since urban rivers were a source of 'disease and death' (Kaika, 2005, p. 6), water in the later nineteenth and large parts of the twentieth centuries, became 'tamed and controlled' (ibid.). The establishment of the link between dirty water and cholera by John Snow in 1845 (ibid., p. 85), made the construction of water supply and sewage networks crucial for promoting the health of urban populations (ibid., pp. 85–86). According to Kaika, such sanitary and hence biopolitical, projects were also pivotal in the construction of the identity of the West, and of Western cities in particular, as they were taken on with 'prestige and pride' (ibid., p. 86). Thus, the construction of water and sewage networks, and the cleanliness they involved, represented a good and orderly city life, compared to the previous era in which disease and death proliferated through uncontrolled water.

Kaika describes how the expansion of urban areas and a new understanding of the link between epidemics and water produced not only a whole 'science, economy and practice of treating and purifying water' (ibid., p. 54), which is characteristically biopolitical (see Gandy, 2014), but also discursive constructions of 'good' and 'bad' water (Kaika, 2005, p. 54). While good water was 'clean, processed, controlled and commodified', bad water was its opposite, that is, 'dirty, grey, metabolized, nonprocessed, noncommodified' (ibid.). In turn, these constructions, simultaneously material and discursive, led to new spaces being created for an interaction between the human body and 'good' water, such as swimming pools and spas (ibid.).

In particular, one of these water spaces, the private house, became *the* place for good and purified water. In contrast, the uncontrolled and unregulated water outside of house, in the public domain of the city, became associated with threat and danger and hence understood as 'bad water' (ibid.). Thus, the ideas of modernity and progress were not confined to the city as a particular space but were also productive of the idea of home. Kaika writes:

> The 19th century idea of the sanitized, piped, wired, plumbed city was to be extended to the idea of the sanitized, piped, wired, plumbed house – the classic icon of 1950s and 1960s advertisements for home durables and home style – which promised, once again, the final delirious satisfaction of our dreams and desires.
>
> (Ibid., p. 41)

Both gender and class relations played a significant role in relation to the different kinds of water and who could enjoy privileges around water. Private bathrooms and new hygienic practices became connected to social status for the middle classes (ibid., p. 55); and while men were taming the 'bad water' and producing 'the good water' through different kinds of infrastructure, women were confined to working with 'the good' water in the domesticated sphere (ibid., pp. 58–59). According to Kaika, there was, however, something else at play; both nature and society were 'othered' in relation to the (Western and middle-class) private home. Such a home was constructed as an 'independent' and 'autonomous' sphere, that is, free from unwanted social and natural elements, such as homeless people, dirt, and pollution (ibid., p. 7). Kaika writes:

> thus, excluding socio-natural processes as 'the other' becomes a prerequisite for the construction of the familiar space of home. The inside becomes safe, familiar, and independent not only by excluding rain,

> cold and pollution, but also through keeping fear anxiety, social upheaval, and inequality outside.
>
> (Ibid., p. 61)

Thus, in this way, modern planning and architecture, through the construction of both water networks and private houses, create a line that separates inside from outside, private from public (ibid., p. 62). As we will see in the next chapter, the dynamics of inside and outside, both in relation to water access and in terms of inclusion and exclusion from the political community, is especially pertinent in the context of informal settlements in South Africa. Here, the lack of proper housing and water access produces a situation in which not all members of the community feel 'human'.

However, even though private and public, nature and culture, are constructed as separate, they are simultaneously constitutive of each other. The outside is necessary as a way of constructing the very familiar feeling of the home both as an abstract idea and in purely material terms. As Kaika notes, a cut-off from water reveals the dependency of the unhidden other (water from 'nature' and the processes of distribution and purification) (ibid., p. 67). In turn, such experiences produce feelings of 'uneasiness, discomfort, and anxiety' (ibid., p. 68) a theme the next chapter returns to.

Even though the hydraulic mission was not one single system of thought, it can – in general terms and on the basis of the above – be described in hydromentality terms as a paradigm that sees water as a resource for building the nation state, its agriculture and industry, and for creating modern citizens. In the case of South Africa, as in other colonial contexts, the citizenship to which ideas of modernity were attached was based on race. In terms of the construction of modern citizens, domestic infrastructure was central, such as kitchens with running water and a private bathroom, including facilities for a bath and a flush toilet. These facilities also aided the process of establishing the notion of the private home, as separate from both 'nature' and public life (Kaika, 2005). Within the frames of this paradigm, water governance was centralized and its decision-making heavily dependent on the technical expertise of engineers. Hence, it governed through top-down decision-making and established roles and identities in relation to water along gender, class, and race lines (for this, see, for example, Kooy and Bakker, 2008a, 2008b). As we learn from critical urban and infrastructure studies (Gandy, 2006; Legg, 2006), the design of cities and their infrastructures can be viewed as a means of governing.

As Linton (2010) indicates, in many cases resource accessibility was taken for granted. However, in a country such as South Africa, because of its relative dryness, water scarcity was already an issue during the period of the hydraulic mission. In a report published in 1970 by the Commission of Inquiry into Water Matters (see RSA, 1970), issues of water efficiency and saving had been discussed versus the construction of new water projects. The hydraulic mission came to be challenged, however, in South Africa and elsewhere.

The hydraulic mission challenged

Critique of the hydraulic mission emerged in terms of its supply-led focus and the vast costs of hydraulic infrastructure (i.e. large dams), along with an inability, or unwillingness, of governments to finance this infrastructure (Bakker, 2010, p. 35; see also Molle *et al.*, 2009). Part of this criticism was also based on the fact that environmental concerns had not been taken into account in the drive to increase water access (Bakker, 2005, 2010, p. 35), especially in large dam constructions (Allan, 2002; Molle *et al.*, 2009). In developing countries, these so-called state failures also involved a failure to provide universal access (Bakker, 2010, p. 87). Due to these factors, water supply management became the target of wide-ranging reforms (for a detailed analysis in a development context, see Hansson, 2013) and paved the way for a new water paradigm, placing greater emphasis on private actors, market solutions, and competition in water governance, as well as on demand management instead of supply-oriented solutions (see Bakker, 2010).

To counteract the diagnosis of state failures, emphasis was placed on processes of privatization, commercialization, and liberalization (ibid., p. 38). Processes of privatization can be understood as a transfer of management or ownership of resources to private individuals or companies and 'the creation of new markets for previously uncommodified resources' (ibid., 38). In a water context, however, it needs to be said that privatization has rarely meant that ownership of water per se has been transferred to private companies. Rather, over past decades, private companies have been participating in the water sector as owners and funders of water infrastructure, as service providers, and through joint ventures or other forms of public–private partnerships (PPP). Commercialization, on the other hand, has involved the inclusion of market actors and market logics, whereas liberalization is a broader concept that includes deregulation, commercialization, and decentralization processes meant to stimulate competition and reduce the role of government (cf. ibid., p. 38). Privatization, liberalization, and decentralization processes of water service provision have worked, in practice, as conditionalities in development cooperation (see Boelens *et al.*, 2015; Hansson, 2013, 2014; this is returned to further on in the chapter). Bakker has termed the new paradigm that incorporates these processes 'market environmentalism' (2010, p. 35).

In this book, the new paradigm that emerged after the hydraulic mission is understood as *neoliberal hydromentality*. While recognizing the importance of the processes of privatization, commercialization, and liberalization for understanding water governance, a governmentality perspective understands the neoliberal character of such governance in a broader sense. As described in the previous chapter, GWG incorporates central components of market environmentalism, such as the importance of the economic valuation of water, and the view that private actors within the water sector are seen as providing an opportunity to create both an efficient and a sustainable management of resource use. However, it does not advocate 'pure' market solutions to water governance, but proposes a range of interventions, including guaranteeing a human right to water and water for the environment, which in the end will make market rationalities function. These interventions should be understood as a form of biopolitics that aims to optimize both human and non-human life. It should here be recognized that the roles of privatization, commodification, and liberalization as (sole) remedies for problems in the water sector, have shifted since the 1990s and, moreover, that the appropriateness of such processes has given rise to tension and negotiation within the IWRM framework. However, a *neoliberal hydromentality perspective* on these deliberations views them as part of the (neo)liberal problematique of governing without governing either 'too much' (Foucault, 2008, p. 319), or too little.

Additionally, addressing water governance as neoliberal hydromentality places a focus on specific governing *techniques*, such as responsibilization and the 'conduct of conduct', that is, the steering of people's behaviour around water. Furthermore, it focuses on how subjectivities are formed in hydropolitical relations of power and how the agency of water users is utilized to secure the objectives of government (cf. Dean, 1999, p. 15). Such techniques of governing are the focus of the next two sections, and their effects are explored in Chapter 5.

Contemporary water governance: neoliberal hydromentality

If water was considered a 'natural fact' in the hydraulic mission paradigm, the way that water is understood has been 'complicated' in contemporary water governance (Linton, 2010, pp. 6, 7, 50). The push for liberalization and privatization, and the fact that the Dublin principles, on which present water governance rests, defined water as an economic good in the fourth principle, have invoked an intense debate on what water *is*.[3] Protesters against water privatization and commercialization around the globe have argued that water should be seen as a right, rather than as a commodity/economic good (Shiva, 2002; for an overview of the debate, see Bakker, 2007). Academics have put forward that water, as essential to all life, is ill suited to the application of market principles (see Barlow and Clarke, 2003; Shiva, 2002). This has spurred a debate on what a right to water entails, both practically and theoretically (see, for example, Bakker, 2007, 2010; Mirosa and Harris, 2012; Morinville and Rodina, 2013; Sultana and Loftus, 2012). In regard to these different ways of trying to pin down the definition of what water is, Bakker has pointed out, however, that water cannot easily be understood in terms of one single identity (Linton 2010, p. 49).[4] Water is 'uncooperative' (Bakker, 2004) and has been far more difficult to commercialize than some would have hoped for and others had feared (Bakker, 2010; Linton, 2010, p. 48).[5] Such a complex understanding of water is characteristic of present governance discourses.

Given the problematization of water in terms of funding for infrastructural projects, scarcity, pollution, distribution, and its relationship to the health of ecosystems, it has become less taken for granted. Water has thus begun to be thought about in a different way (Linton, 2010, p. 6). Linton argues that IWRM can be understood as an effort to 'reconcile the growing awareness of water's fundamental complexity with its rather one-dimensional identity as a resource' (Linton, 2010, p. 50). Water is not just a resource, it has ecological, cultural, and political dimensions (ibid., p. 7).

What this also means is that water problems, such as scarcity, can no longer be naturalized as in the previous (hydraulic mission) paradigm. Hence, water access is no longer (only) a question of making it available through hydraulic technologies, but is related to ideas of (good) governance, as outlined in the previous chapter. For Linton, this does not mean that the IWRM approach challenges the underlying assumption of 'modern water' (see especially Linton, 2010, chapter 11), but that the way the global water crisis was constructed in the 1990s, which subsequently gave rise to a new water paradigm, was in fact framed within the terms of modern water (ibid., p. 93). According to Linton, the crisis was constructed through a description of an unbalanced relationship between the availability of water, presented in abstract terms, and a similarly abstract notion of people, presented in their 'statistical guise as "population"' (ibid., p. 193). In this way, the water crisis was presented as something static that did not, in any deep sense, challenge the paradigm of modern water. Challenging this paradigm, Linton argues, would involve changing the relationship to water and its use. Ironically, the way that the water crisis is presented works against the realization of a new water paradigm. Linton states: 'Modern water does not constitute a crisis all by itself; it requires modern people' (ibid., p. 197). As has been argued in this book, in water governance and in sustainable development discourse in general, something additional is at stake as regards the notion of population. Not only is the population in these discourses related to the availability of resources, but when governance enters into issues of resource use and environmental care, it also becomes a question of how different populations are understood in terms of their relation to futurity.

Furthermore, the conceptualization of the different 'values' of water in GWG is instructive in terms of how this governance approach deals with the questions of what water *is* and how we are to understand its *resourcefulness*.

It is safe to say that discussions around the 'values of water' are a matter of controversy and a sensitive issue because they invite critique against market-based solutions in general and against water pricing in particular, which are two important components of GWG. Furthermore, the understanding of the values of water varies greatly between different contexts.

At present, there are discussions within the frames of GWG on the need to consider the multiple values of water and how to incorporate these different values into decision-making. The work is carried out by a High Level Panel on Water (HLPW) and is facilitated by the Global Water Partnership. A concept paper, based on the Bellagio principles of water, has been prepared as a discussion document for consultation between different stakeholder groups (including NGOs/CSOs, indigenous peoples, faith-based groups, governments, academia, youth, women, and private sector organizations) in selected countries (GWP, 2017).[6] The preamble states:

> **Water is precious, fragile, and dangerous.** It can sustain or destroy. Water in combination with land, air, and energy is the foundation of life, societies and economies. Water is more than a substance. It carries multiple values and meanings. These are expressed in spiritual, cultural and emotional terms and found in the heritage of water language, norms and artefacts. These reflect the deep perceptions, need for connections and participation of all of society. Making water available for its many uses and users requires tools and institutions to transform it from a natural resource to one providing services and then to recover and return it safely back to nature. Water and its sources must be respected, because if neglected it has the power to harm, divide or even destroy societies.
>
> (Ibid., bold text in original)

Furthermore, it recognizes that '[m]aking all the values of water explicit gives recognition and a voice to dimensions that are easily overlooked' (ibid.).

Such efforts at incorporating a multiplicity of viewpoints regarding the value of water can be contrasted with how water's value is discussed in the WWDR reports; a quick search through the last three WWDR reports (WWDR, 2015, 2016, 2017) reveals that the word 'value' is most often connected to *economic* value, even though the value of water for both social purposes and ecosystems is acknowledged. However, the way in which the latter two is considered often ends up producing an economic argument. An understanding of the relationship between the environmental and economic dimensions of water is instructive here. The environmental dimension of water is often related to ecosystem services and EBM is seen as a tool that can safeguard long-term sustainability of water use. From such a perspective, the problems with conventional economic models are that freshwater ecosystems are not properly valued for the services they provide, and furthermore, that untreated water from residential areas, industry and agriculture weakens the capacity of such ecosystems. While acknowledging ecosystem functioning as central to water governance, the well-being of ecosystems is often connected to the economic benefits that they produce and the cost of replacing them. Ecosystem services and EBM thus involve an inherently economic perspective on what water is and does.

How water is valued in different contexts also relates to how the incorporation of global governance reforms in national water policy and practice plays out in different national and local contexts. As we shall learn from those water studies that have applied a governmentality framework presented below, in many places this has involved an 'othering' of hydromentalities different from the dominant understanding.

Not only have scarcity and pollution changed and complicated how water is conceptualized within the present water governance paradigm, but climate change has also produced new ways of understanding water. Water has increasingly come to be seen as both uncertain and as posing a threat (even if water, or the lack of it, has always had the ability to create destruction through, for example, drought and flood). In contemporary GWG, water is thus conceptualized as *dangerous*; note, for example, the preamble to the Bellagio principles: 'Water is precious, fragile, and dangerous'. Such a conceptualization of danger and risks is related to the water security concept.

In sum, the present water governance paradigm can be described as a neoliberal hydromentality that (drawing on the previous chapter) defines water as a resource with multiple identities and uses. However, it is first and foremost understood as a resource that produces economic value determined by the mechanisms of the market. The guaranteeing of basic human and ecosystem needs is, however, seen as entailing necessary interventions so that 'competition and enterprise' as regulatory mechanisms 'will be allowed to function in the rest of society' (Foucault, 2008, p. 206). In this paradigm, water scarcity, pollution, and the lack of basic services, as well as uncertainty and risk following climate change, are understood as the main challenges. The solution to these challenges, as they appear in policy, is not only holistic, inclusive, and gender-sensitive governance, but also technological measures, such as desalination and reuse of waste water, and mental and material resilience and preparedness. These policy processes simultaneously involve the drawing of lines both between different populations – in terms of their relation to resource access, use, and sustainability – and between human and non-human life, a distinction made so as to construct the latter as a resource base for the former. Techniques that are promoted in order to govern water use and water users within this paradigm, include participation and consultation, demand and ecosystems-based management, and the use of technological devices in water service provision. These techniques are employed so as to responsibilize water users, and to steer their behaviour and ways of understanding themselves, as is empirically explored below and in the next chapter.

Hydromentalities: experiences from across the globe

One of the most important points of the methodological framework that guides this book is that we need to make careful empirical investigations of water governance in order to grasp the impact of (global) water governance regimes. A few contributions that have explicitly applied governmentality theories to the study of water are discussed below. These studies provide perspectives on different aspects of particular hydromentalities. Together, they give a few general insights both into past and present water governance regimes and into what a governmentality perspective adds to the understanding of the politics of water.

Technologies of rule, responsibilization, and resistance

A study of the governmentality of water reform in Ecuador by Boelens *et al.* (2015) is an instructive example of the importance of studying the particularities of specific cases in relation to global governance prescriptions. The case is enlightening in the sense that, through its governmentality perspective, it shows how the reform of water policy not only aims at governing water issues but is also part of a larger process of reshaping the relations between state, market, and society. It also provides an empirical example that reflects how, over time, water governance discourse has moved away from market-based solutions to a more integrated and inclusive approach, as formulated in policy documents, at the same time remaining within a neoliberal framework both as regards economic policy and in relation to the governing techniques employed.

In Ecuador a new neoliberal water management model was crafted in the aftermath of the economic crisis and experiences of corruption and inefficiencies in Ecuadorian state bureaucracies in the 1990s (Boelens *et al.*, 2015, based on Hey and Klak, 1999). This shift was supported by the World Bank, the International Monetary Fund, and the Inter-American Development Bank, and meant that Ecuador could get access to new loans from the World Bank on condition that the country opened up the water sector to private sector involvement (Boelens *et al.*, 2015).

The same development took place in other Latin American countries. Typically, water reforms involved processes of privatization, commodification, marketization, reduction of state involvement, and the deregulation and decentralization of water management (ibid.; see also Ward, 2013, discussed below). In Ecuador this meant that water shifted from being seen as a public good, owned and governed by the state, to being conceptualized as a 'scarce commodity, the services of and access to which had to be allocated to consumers (or "clients") by means of market mechanisms' (Boelens *et al.*, 2015, p. 284). The direct involvement of state institutions in water affairs was reduced. Instead, the state was focused on 'making the market function' (ibid., building on Cremers *et al.*, 2005). The prevailing rationality was that market forces and the private sector would facilitate efficiency in water management. In this process, water user organizations, mixed public–private partnerships, or private companies, were given more space (ibid.).

In the case of Ecuador, the result of these wide-ranging reforms was fragmented water governance. Water user organizations started to manage their own irrigation and supply systems. State authority became contested by water user organizations, private entities, and local government, who had assumed new roles (Boelens *et al.*, 2015, p. 285, building on Hoogesteger, 2012). This led to more powerful actors being able to strengthen their influence. Local water users were granted more autonomy and control over water systems. At the same time, peasant and indigenous communities experienced problems due to the reduction of state support for irrigation and agriculture. Water agencies that were created in the course of the reforms were weak and unable to handle conflicts in relation to water use. This led to everyday water problems and to more long-term effects on power and authority in water governance. As a result, environmentalists, water user organizations, and indigenous federations advocated for a replacement of the neoliberal policies that had swept through the country.

Rafael Correa won the 2006 election, promising to replace neoliberal policies. In relation to water issues, his *revolución ciudadana* (citizen revolution) meant that natural resources were re-nationalized, which granted the state, once again, control over water and water supply. This means that expenditure on water administration, hydropower, and irrigation projects was raised (Boelens *et al.*, 2015, p. 285). The state strengthened its position through the National Water Secretariat (Secretaría Nacional del Agua, SENAGUA), which took over the responsibility of coordinating national water management through a watershed-based, integrated water management framework (Boelens *et al.*, 2015).

Correa's transformation of water governance emphasized 'equality, inclusion, participation and democracy' (ibid., p. 286), while at the same time both criticizing and incorporating the neoliberal economic model. The authors write that:

> A key feature of the citizen revolution is the attempt to re-build a strong State, re-gaining its regulatory and enforcement capacities and becoming the sole authority for managing society and its economy. Discursively, the citizen revolution strongly attacks the neoliberal Economic model but its practices support many elements of the same model.
>
> (Ibid., pp. 286–287)

Moreover, the Ecuadorian government uses techniques of governing associated with neoliberalism as an art of government (Dean, 1999, p. 210), techniques such as stimulating the acceptance of the framework through the training of water users. The water users are thereby subjected to performing the 'responsibilities of well-trained, compliant water users' in accordance with 'the rules on efficient water use and modern governance' (Boelens *et al.*, 2015, p. 285). Relatedly, participation and inclusion in decision-making regarding water issues have taken a particular shape in Ecuador. The authors write that collectives are only allowed to take part in decision-making as long as it is '*in the interest of the state*' (ibid., italics in original) and the national revolution. The result is that influence from civil society, including unions, grassroots organizations, and peasant and indigenous federations, has been reduced, but that business entrepreneurs within the government sphere have strengthened theirs (Boelens *et al.*, 2015, building on Ospina-Peralta, 2009, p. 12).

Hence, official policies that embrace participation and inclusion are not enough to produce progressive water governance. In the Ecuadorian study, the authors argue that water users need to organize in order to claim the people's rights and a more just and democratic water distribution and governance. In this way, the authors claim, 'bottom-up' and 'grassroots-driven' initiatives can work as a 'strategic resistance' to both a state-centred and a market-centred water governance (ibid., p. 289).

Similar dynamics to those in Ecuador, with regard to using people's capacities and agency in strategies of governing, have also been observed by governmentality scholars focusing on water governance in India. Babu's (2009) study of rural drinking water supply systems in Kerala provides perspectives on how participation is used as a governing technique, while Birkenholtz (2009) addresses the construction of subjectivities and self-monitoring water users in Rajasthan's groundwater governance.

Babu argues that participation and awareness programmes in the local context of Kerala are used as a way of meeting the ends of the state and softening critical voices over water service delivery. Empirically, the study takes the example of Jalanidhi, a rural water-supply programme supported by the World Bank, which was initiated in 1998. Babu explains that people's participation has been a principal strategy in regard to water supply since the 1990s in Kerala. This is due to the idea of decentralization that became the response to the failure of 'big government' to provide basic services such as water. Kerala has, furthermore, been acknowledged as an exemplary case in relation to participatory strategies (Babu, 2009, p. 92).

The Jalanidhi project is built on a model in which some of its capital costs and the maintenance are borne by the beneficiary communities. Water is provided through the grid by a collaboration between different actors (the Kerala Rural Water Supply and Sanitation Agency, local self-government institutions, and an NGO). Responsibility for planning and implementation lies at local community level. This is celebrated as a successful alternative to both state-run and privately-run utilities. However, according to the study, a closer look at the dynamics of the project reveals that the model is merely what Babu calls 'a staged drama ... created and shaped deliberatively by the state' (ibid., p. 94).

To stimulate the active participation of community members, NGOs and experts (social workers and technicians) use techniques such as Participatory Rural Appraisal (PRA) in awareness programmes. Babu argues that discussion within such awareness programmes is steered towards the acknowledgement of the (lack of) responsibility of community members in dealing with problems associated with water access and towards the notion of inefficient government institutions.

In this system the individual (in collaboration with the community) becomes responsible for securing their own water needs through addressing water issues in user groups, consisting of about 10–20 households. Each user group defines its own rules and responsibilities and has a president, secretary, and treasurer who have been trained, by NGOs, in bookkeeping, accountancy, and so on. Decisions in the user groups are then made by these key people, together with NGO representatives and officials. Even though these people have been elected by majority decision, Babu argues that, in practice, only those with economic, cultural, or political capital can be elected, leaving others without a voice in the decision-making process (ibid., p. 94).

In sum, what Babu argues is that the participation discourse in water management serves the purpose of responsibilizing local communities and individuals and internalizing global discourses around issues such as hygiene and payment for water, while taking the edge off criticism of the state.

Birkenholtz' study (2009) of groundwater governance in Rajasthan takes place in a context characterized by tension and even eruptions of violence against protesters voicing their concern over insufficient access to water. The study focuses on decentralization processes and outreach programmes designed in line with global environmental governance and explores the resistance to such ways of governing. The methodological set-up of the study integrates insights from both neo-Gramscian scholarship, through the concept of hegemony, and green governmentality literature and its focus on the intersection of environmental governance and the making of subjectivities.

Birkenholtz describes the development if groundwater governance in Rajasthan towards one that is neoliberal in character, carried out with the assistance of the World Bank and other development agencies. These reforms include decentralization processes, the deployment of awareness programmes, a focus on establishing property rights, and the introduction of technologies meant to increase efficiency. This is also a story of how the state, with the assistance of donors, NGOs, and the World Bank, has tried to formalize water regulation where previously there had been only informal governance (ibid., p. 212).

According to Birkenholtz, decentralization, as it operates in the case of Rajasthan, rather than decentralizing decision-making to the local level, serves to devolve state control. This is done in two ways. Representatives to local water governing bodies are appointed so as to merge the interests of local elites (including agrarian upper caste/class elites and local state officials) with those of the state. Support is also sought through 'outreach activities' organized in order to change the attitudes of farmers who are highly dependent on groundwater sources (ibid., p. 217).

Whether farmers oppose or support reforms is, however, connected to whether they believe that they will gain or lose, according to the findings of the study. As this, in turn, is related to local hierarchies, caste and class become central to the way they relate to the environment and environmental governance. Birkenholtz notes (ibid., p. 218) that these conclusions stand in contrast to those of Agrawal's study (2005) on forest governance in India (discussed in Chapter 2), where the assessment was that such factors did not play a significant role.

In Birkenholtz' study, resistance is one of the key themes and a number of tactics and activities are utilized to resist and protest against reforms: protest marches; nonparticipation in outreach programmes; and the forming of informal irrigation partnerships, which are particularly aimed at resisting efforts to implement individual property rights over groundwater. Analysing resistance is of the utmost importance when attempting to make sense of the dynamics in hydromentalities, and this is also a theme in Boelens *et al.*, (2015), Kooy and Bakker (2008), and in the study of the effects of the hydromentality of eThekwini municipality, discussed in the following chapter. Departing from a governmentality perspective, which emphasizes 'subtle' techniques of governing (Birkenholtz, 2009, p. 209) – such as working on people's subjectivities so they can be self-governing, as exemplified in several of the studies referenced here – studying resistance involves certain methodological challenges, which are discussed in the next chapter.

Subjectivities and agency of water professionals

Another research focus of governmentality studies of water governance is how water professionals navigate their roles in governance shifts. One example of this is Luke Ward's study of the subjectivities of IWRM experts in Paraguay (Ward, 2013), which focuses on the uncertainty of the outcomes of transformations of (neoliberal) governmental regimes.

In Paraguay the transformation to an IWRM approach involved the introduction of new rules and planning regimes. In 2002 the World Bank, the Global Environmental Facility (GEF), and the UNDP initiated IWRM-based projects. One of the core activities has been to train state, NGO, and civil society actors to take part in development and environmental planning in IWRM theory and methods. Ward examines how IWRM logics translate to these water experts in terms of how they cope with policy and planning in relation to the 2007 IWRM water law and processes of development of environmental zoning plans for watersheds (POATs) (ibid., p. 92).

Through this study we learn that while the participants share the goals of environmental and IWRM governance, that is, poverty reduction, empowerment, and public participation, the same actors disagree about how these goals should be achieved; particularly in terms of the roles for different actors, be they non-governmental organizations, the state, or local actors, as well as regarding the role of the market. Ward builds on previous literature (Bakker, 2007; Ong, 2007) when he concludes that a study such as this reminds us that governmentalities, like those in IWRM, do not materialize in a straightforward way when they are introduced. Rather, such processes not only involve a commitment to reaching the end goals of governance, but also entail a strategic engagement of different individuals representing their own and other actors' importance and political agendas (Ward 2013, p. 100).

Another study that focuses on professionals in water management is Stina Hansson's work on water sector reform in Niger (2014), especially focusing on the process of achieving Nigerien ownership of the water sector through the implementation of the programme approach. The study uses a governmentality framework combined with narrative and postcolonial literature when analysing data consisting of state agent narratives and their participation in ministry–donor negotiation meetings and workshops.

The case of Niger is particularly interesting given its high dependence on aid (80–90 per cent of the water budget) and the fact that the country has followed global development and water governance trends. Following these trends, there was a push for privatization and decentralization in the 1990s. Aid has been reorganized into result-based managed programmes. The reforms have involved a delegation of responsibility to private and local actors in the rural sector and to PPPs in urban and semi-urban sectors. Hansson notes that while these reforms are still central to water sector reform, the focus on an effective state has been re-emphasized. In this context, the programme approach has become a tool for improving Nigerien state institutions' capacity to govern the water sector (ibid., p. 259).

Hansson builds on previous literature on ownership (see, for example, Abrahamsen, 2004), when arguing that the programme approach, as it is being implemented in the Nigerien water sector, is an 'illustrative example of how ownership works through technologies of agency and performance' (ibid., p. 272). The space within which the Nigerien state can act is small, given that reforms such as decentralization, privatization, and results-based management are pre-defined as necessary to global development and water discourses. The way that the active agent (i.e. the Nigerien state), produced in these relations of power understands and makes use of the space to act, nevertheless shapes the way that ownership logics play out in the specific context of Niger.

The study paints a complex picture of what ownership means in the context of the Nigerien water sector. In the narratives of the state agents, the possibility of ownership is understood in relation to 'the exclusionary practices of aid in the past (and present) as promising and, to some extent delivering, inclusion and capacity to act in the water sector in the future' (ibid.).

The formalization of a donor presence in decision-making is interpreted by state agents as providing a possibility to imagine an active state; at the same time, however, such cooperation is viewed as a site for compromise. Furthermore, state agents do not see a contradiction in calling policies that have been formulated with northern consultants (in order to guarantee donor support) 'their own'. The study indicates that in this context, ownership has come to mean the 'capacity to act' rather than denoting a state of self-determination (ibid.). Hansson writes that:

> The state agents in their marginal position do not imagine autonomous decision-making; rather, their main objective amounts to being included in the process of developing. To imagine the state as provider of services at all, donor commitment must be safeguarded as it provides the working conditions of the state.
>
> (Ibid., p. 273)

Important here, however, is that this way of conceptualizing ownership should not be seen as 'mere compliance'. It should, rather, be understood as a way for state agents to find a position from which they can act (ibid., p. 273). Thus, as in the above discussed studies, we learn how the state is constructed in relation to global water governance, and more broadly in relation to global development policies and discourse, through neoliberal techniques of governing.

Infrastructure, geographies of difference and identities

A third major theme that can be identified in the water governmentality literature is infrastructure and its role in constructing geographies of difference and identities.

A study by Rogers *et al.* (2016) analyses, from a governmentality perspective, China's South–North Water Transfer (SNWT). The project is the world's largest interbasin water transfer project and is unprecedented considering the vast scale of the four major river basins that are to be connected. The study explores the project in terms of its 'machinery, mentality and spatiality', its 'constitution of objects and subjects', its techniques of government, and 'its physical and administrative assemblages' (ibid., p. 429). It does so through analysing English and Chinese academic, media, and government documents in terms of Dean's (2010) four dimensions of governmentality: forms of visibility; techne of government; episteme of government; and forms of identification.

The authors describe how a particular imaginary of a seemingly evident need for taking water from the water-rich south to the water-scarce north, to compensate for geographical differences, is created within the frames of the project and, furthermore, how the project prioritizes particular forms of knowledge that promote engineering solutions. By doing so, the project naturalizes and depoliticizes both water scarcity and the SNWT project, while simultaneously prioritizing the Beijing area.

However, governmentalities are not always successful or consistent (see Rutherford, 2007; see also Hansson *et al.*, 2015). The authors describe a project whose infrastructural and institutional set-ups are only 'partially effective' in administering and governing water. What we learn from the study is that China does not operate like a hydraulic society à la Wittfogel (1957)[7]. Rather, what the authors depict is that Chinese (water) governmentality has multiple centres and is one that merges market interventionist logics and techniques in an assemblage of actors and relations that is not fully controlled or controllable (Rogers *et al.*, 2016, p. 437).

Perhaps one of the most well-known studies of water and water infrastructure to apply a governmentality perspective in a particular context is Kooy and Bakker's study (2008) of the relationship between environmentality and materiality in urban space in Jakarta, as discussed in Chapter 1.

Through a framework that focuses on the 'relationship between governmentality, identity, urban space and urban infrastructure', the article explores differentiation in water supply and how it has persisted from the colonial period through programmes of development and neoliberal governance. The study finds that the way the 'modern' urban citizen has been discursively constructed is related to divisions along racial lines established in the colonial era, divisions which have later been reinforced along with socio-economic markers. Thus, the study finds that access to piped water, or the exclusion from it, has been dependent since the colonial era upon one's identity, either as 'modern' or 'in need of development' (ibid., p. 376).

The study is an important contribution to the wider literature on (post) colonial urban spaces and water politics in the Global South. Three points are noteworthy here. First, the study points to the centrality of binaries – 'European/native, modern/primitive, civilized/uncivilized, sanitary/insalubrious, developed/backward' (ibid., p. 385) – in the technologies of government in both colonization and modernization projects. Second, the article makes the case that it is not possible to understand processes of exclusion from water access unless 'materializing' the analysis of technologies of government, hence focusing on the 'mutual reinforcement of material, spatial, and discursive aspects of "technologies of government"' (ibid., p. 386). Third, the study acknowledges resistance as part of how technologies of governing water infrastructure are unfolding.

The role of infrastructure, and how it differentiates between populations and their part in the construction of subjectivities, is also a theme in the research on the hydromentality of eThekwini municipality outlined below.

The hydromentality of eThekwini, South Africa

My own research (see Hellberg, 2015, 2014, 2017) has described the hydromentality of one specific municipality: eThekwini. This particular hydromentality has been produced in the context of the South African constitution and water legislation created after the end of apartheid (RSA, 1996, 1997, 1998). This new legislation became seen as one of the most progressive water regulations in the world in terms of its integration of the principles of IWRM and its emphasis on coordinated, equitable, sustainable, and economically viable water management (Ashton and Haasbroek, 2002, p. 191; Mirumachi and van Wyk, 2010; Pollard and du Toit, 2011). As already indicated in the introduction to this book, the fact that the South African constitution defines 'sufficient water' as a right has gained global attention.

It was in eThekwini that the FBW policy was developed. The case of eThekwini is an exemplary one in terms of the implementation of a FBW policy and in relation to its innovative policies and technologies. In essence, the FBW policy means that a minimum of 25 litres of potable water per person per day should be available within 200 metres of a household (RSA, Government Gazette, 2001) and was given legal status in 2001. The municipality is thus a highly relevant case to study as it is exemplary in relation to global norms of water management (see for example the booklet: *Most Innovative and Progressive water utility in Africa: Selected Highlights* [Water and Sanitation, n.d.]). EWS has received the UN's public service award (EWS, n.d.), and 'A Water for Life best practices award', acknowledging its efforts in 'raising awareness about water conservation and improving sanitation' (WRC South Africa, 2011). In 2013 Neil Macleod won the IWA Development award for pioneering 'innovative approaches to better serve poor communities', and for fostering 'partnerships between stakeholders and service providers within his municipality and also across Africa' (IWA, 2013). Furthermore, in 2014 the EWS was awarded the Stockholm Industry Water Award (SIWA). The jury supported their choice in the following words:

> eThekwini has championed the approach to provide sufficient water to sustain human life, as expressed in the South African constitution, now embedded in national policy. The methods used and results achieved by eThekwini Water and Sanitation serve as a sterling example for the many communities worldwide facing similar challenges.
>
> (SIWI, 2014)

However, the progressiveness of the case of eThekwini is contested. It has been the target of criticism from academics and civil society alike, directed at tariff and payment systems, at the practice of disconnecting residents from water access, and at the inadequacies of water systems in rural areas

(Bond, 2010; Loftus, 2005a, 2005b; see also Galvin, 2016; Loftus, 2006, 2007, 2009; Narsiah, 2010; Nash, 2013; Sutherland *et al.*, 2015).

eThekwini municipality was created in 2000 through a number of boundary re-demarcations. It consists of the city of Durban, surrounding peri-urban residential areas and rural communities. According to the 2006 integrated development plan (IDP), the spatial configuration of the area has resulted in 'a distinct pattern of inequity and inefficiencies across the municipality' (IDP, 2006/2007, p. 97, see also Freund, 2010; Freund and Padayachee, 2002 for detailed descriptions on the social and economic geography of Durban).[8] In terms of service delivery, including water, the communities that are worse off are those in the historically under-invested in township areas, including informal settlements and in the rural periphery (IDP, 2006/2007, p. 98). eThekwini faces vast challenges in terms of high levels of unemployment, high levels of poverty, low levels of skills in the work force, high levels of HIV/AIDS and, as in every major South African city, high levels of crime (IDP, 2006/2007, pp. 9, 101). These challenges also remained in the 2016/2017 IDP. In terms of rolling out basic water services, the 2016/2017 IDP however notes that 'tremendous progress' has been made (IDP, 2016/2017, p. 11). The following provides an overview of the kind of hydromentality that produced these outcomes. This description is based on interviews with key people at eThekwini Water and Sanitation (EWS), an analysis of policy documents, and on observations. The fieldwork was conducted between 2009 and 2010. Chapter 5 explores the biopolitical effects of the hydromentality of eThekwini from the perspective of water user narratives.

A regulatory framework

The hydromentality of eThekwini is characterized by the idea of water as a basic right and as a commodity, and with sustainable development as a guiding concept. Sustainable development is understood in relation to the three dimensions commonly associated with the mainstream approach: social, economic, and environmental. This means that the municipality judges solutions against whether water services are affordable, whether the funding and financing will last for generations, whether society will accept, use, and value them for generations, and whether the impact on the environment is neutral or positive (Macleod, 2009).

EWS is known for being innovative in terms of creating new policies and technologies. It was in eThekwini municipality that the policy of free basic water, which later became a national policy, was developed. At first, the policy meant that each household received 6,000 litres monthly.[9] This is equivalent to 25 lcpd when calculating that a household consists of eight members. The FBW amount was raised to 9,000 litres in July 2008. When it comes to determining the amount that should be provided for free, the rationality is a matter of economic calculation in relation to people's needs.

According to the municipality, the amount needs to be sufficiently high for people not to get sick and to limit the tampering with meters and devices (that is, constructing illegal connections), as this produces costs. It is also necessary for FBW to be cross-subsidized by high-volume end users (Macleod, 2009).

Such an economic rationality in relation to the right to water was also clearly demonstrated in the final verdict in a case over the right to water and prepaid water meters in Johannesburg that went all the way to the constitutional court. In this case, the prepayment meters were challenged in court by five residents in Phiri. The applicants argued that the automatic disconnection of the meters was a threat to security and health, as well as to dignity, and that the FBW household allocation of 6 kl per household per month was insufficient to meet basic needs. The applicants demanded 50 litres of free basic water per day (compared to the standard 25 litres) (Dugard, 2007).

The case elucidates the underlying rationalities of the South African water governance framework; it becomes clear that the law, rather than fixing in detail how much water the municipalities are liable to provide to their citizens, has a *regulatory function*. The following quotation from the constitutional court is indicative:

> [f]ixing a quantified content might, in a rigid and counter-productive manner, prevent an analysis of context. The concept of reasonableness places context at the centre of the enquiry and permits an assessment of context to determine whether a government programme is indeed reasonable.
>
> (ZACC, 2009, p. 30)

According to this logic, the implementation of the right to water is dependent on an (economic) analysis of what is 'reasonable measures', within the state's 'available resources'.

This can be understood as a general characteristic of neoliberal governmentality in which '[t]he economic question is always to be posed within the field of governmental practices, not in terms of *what may found it by right*, but in terms of its *effect*' (Foucault 2003, p. 15, my italics). Instead of asking what the citizens of South Africa are entitled to, the question becomes: 'What is economically rational?' Furthermore, asking 'what is the lowest amount of water that people can rely on without getting sick?' can be understood as a biopolitical rationale that runs contrary to a management approach that aims at redistribution.

As has been outlined previously in this book, the distinction made between those who rely on basic services and those who can afford to pay for higher levels involves a *biopolitical division*. Furthermore, technologies have been developed and applied to certain geographical areas or different populations and users. Such a practice is central to the hydromentality of eThekwini. The techniques through which the water users are governed include both material technologies, such as meters, and technologies that focus on creating awareness, changing perceptions, and ultimately creating self-governing water users. These techniques are outlined below.

Governing technologies: artefacts and policies

The eThekwini municipality has adopted three levels of services offered to domestic customers. Poor rural areas have been provided with the *ground tank system*, which guarantees that FBW is fed daily into a tank in the yard, within the realms of the water and sanitation programme. Besides water, enrolment in the programme also means that each individual household is provided with a urine diversion toilet (UD), which is a dry sanitation solution. The other two service levels are the *semi pressure supply* received by the household via a roof-tank, and the *full pressure water supply* fed directly to the household from the supply network. Domestic customers are charged different tariffs based on the type of water service they receive and the amount of water they use. These tariffs are progressive, with prices rising for each price block (EMWP, 2008/2009; WSDP, 2004).

Another water management device is the flow limiter, which is a meter that can be set to restrict water supply to certain amounts. It can be set to allow FBW only at higher levels, or, it can allow a flow of an unrestricted amount of water. In informal settlements, water dispensers (or standpipes, as they will be referred to henceforth) are installed as an interim measure and, as with the ground tanks, the water is free of charge (EMWP, 2008/2009).

If a water bill is not paid in full by the due date, fees are added on to the outstanding amount. A disconnection policy following non-payment allows the council to restrict the consumers' water using a specially designed washer, or trickler, that only allows water to flow through at an extremely low rate at a daily consumption of approximately 360 litres in a six-hour period. Water users who have not paid for water for 60 days and owe more than ZAR110 to the municipality will have a flow limiter installed.[10] If flow restrictors are tampered with on two occasions the complete water connection is removed (ibid.).

One of the measures for coming to terms with non-payment is the 'debt relief scheme'. This gives a person who has accumulated arrears an opportunity to write them off by signing a contract with the council, which enjoins the customer to pay his or her water bill on time and in full over a period of 20 months. If they fail to pay, he or she will have additional time to manage the write-off period, and if the unpaid accounts exceed four months, the contract will be terminated. The customer has a choice, either to stay on normal supply or to have a flow limiter installed and thus limit consumption to 300 litres a day. The rationale for this, according to the policy document, is 'to change the mindset of the customers in terms of encouraging them to pay the current charges whilst having their arrears written-off' (ibid., p. 6).

In conclusion, developing and applying different levels of service and technological devices, such as those described above (ground tanks in rural areas, flow restrictors in township areas where people have accumulated debts, regular meters in the townships and suburbs, and so on), appears to be a strategy of governance that distinguishes between different populations and individuals. In this case, it works so as to differentiate between those who will have to live within the limits of basic needs and those who can afford to pay for a higher standard of living; between legal and illegal consumers; between those who live in (rural) areas where infrastructure is too costly to invest in, the so-called the waterborne edge (Gounden *et al.*, 2006), and those who live in urban areas; between those who are informal dwellers, and therefore not eligible for individual access points, and those living in formal housing.

Water scarcity and different populations

Divisions such as those described above also come to the surface in relation to water scarcity. The general message from the EWS was that too much water was being used in the municipality, as an example the following pamphlet was distributed to its 'customers':

> Without water there would be no life! That is why there is a problem. We're using TOO MUCH water! We have almost the same amount of rain each year – but our population keeps growing! (Capital letters in original)

This pamphlet promotes the idea that the problem of water is a straightforward one, and that there is an equation between the quantity of water resources and population size. Other pamphlets appeal to water users to become sustainable and responsible, with pleas such as 'Make every drop count!' and 'Don't Waste Water – Use it!'

The simplified messages obscure the more complex relationship between different water use(r)s and water availability. In an investigation of the relationship between financial sustainability and resource scarcity in relation to water consumption in the richer suburbs, another rationality became evident. In those areas, consumers who use a high volume of water pay more for their water. This provides EWS with possibilities to cross-subsidize. Hence, the relatively large consumption of water in the richer suburbs is economically beneficial to the system. On the other side of the coin, there are economic incentives to keep consumption low (but high enough to sustain survival and health) in the poorer households. Thus, according to this way of looking at the relationship between use and available resources and the problem of resource scarcity, 'wasted water' can be seen as *water that is not being paid for*, rather than the use of high quantities of water. Relatedly, what we see here is that a distinction is made between different populations in terms of what is seen as sustainable water usage.

The issues of wasted water and illegal connections (also seen as a central problem) are good examples of how the hydromentality of eThekwini relates directly to a neoliberal rationality, not only in terms of its economistic ways of reasoning but in terms of its views on biopolitics. Macleod asks:

> How do we deal with people who steal water? How do we reduce the wastage? How do we get cultural change? How do we get people to value water and realise that they have the right to sustain basic life but the rest they must treasure? And if their neighbours are stealing, they must report it to us ... if their water goes off, then it's because somebody has messed up the system by using excessive water ... *if they constrain themselves to what the system was designed for everybody would have water* and we wouldn't have to build the next dam. The wastage is driving that, so cut the wastage! And the major contributors are illegal connections and theft/by-passing.... And then *we need people to feel that's it's my social responsibility to report these things* and to get them corrected ... *not to punish but for sustainability's sake.*
>
> (Interview with N. Macleod, 2009, my italics)

Macleod's statement is telling in the way it forecloses other ways of imagining service delivery. It assumes that a line between water for basic purposes and other uses can, and should, be drawn. It further presents the design of the water service delivery system as socially neutral and implies that deviant behaviour should be understood by everyone as illegal and should be reported and corrected. In other words, it is assumed that the system works in the everybody's interests, which is not how it is understood by (some of) the water users themselves, as explored further in the next chapter. Moreover, Macleod's statement argues for an informer culture in the communities and for a view of those who connect illegally to the network as water users who are spoiling the system for everybody else. Hence, in Macleod's view, water users should conduct themselves so as to 'better contribute to the overall life of the population' (Alt, 2016, p. 96).

The perception of perceptions

As indicated in Macleod's interview, there is a sentiment within the EWS management that perceptions are a vast challenge in relation to service delivery. Macleod spoke of this in terms of a discrepancy between 'political' promises and 'reality'. He explained that politicians and political candidates, in their mission to gain power, promise things that cannot be delivered, and expanded on this topic:

> the concept [that] everybody will have a 300 square-metre brick house with bathrooms and showers and flushing toilets and unlimited electricity and water. That's not a reality, not anywhere in the world. It takes time. It took Europe a thousand years to get there. These guys promise it in one year 'before I leave in four years you will have this' and then they turn to you [as the service provider] and the engineers [and they say]: 'you deliver', [we say]: 'have you got all the money? What about sustainability? What about affordability?' Oh no, that's what they promise. So you have these expectations that are there, and the reality here. And it's hard, hard to improve people's lives in a rate that both we and they can afford, and to moderate their expectations.
>
> (Interview with N. Macleod, 2009)

Hence, according to Macleod's storyline, the challenge of improving people's lives also includes the task of 'moderating' perceptions of what can be expected. In this way, 'reality' is something that is fixed in relation to financial resources and notions of sustainability, as these parameters set the limit for the development of service delivery.

However, not only does this way of producing an objective reality in terms of the different constraints prescribe what can be delivered, but it can also be used to assess the validity of the water users' perceptions. Singh explained his surprise over a report that had been written about water users' perceptions of their services in different areas:

> in the areas where I thought we knew we were not doing very well – based on the actual results – perceptions on the ground: people were very happy with their services. In other areas where I thought we were doing very well, people said you are doing absolutely badly! So perceptions are not, they are not based on realities. It is a perception. So based on your reality, facts and figures *you do need to correct perceptions.*
>
> (Interview with M. Singh, 2008, my italics)

Through their ways of speaking about this, Macleod and Singh discursively moved away from the transformative and redistributive potential of water and the perspective that a fairer distribution of water access can transform society to a more just and equitable one. Instead, they adopted a rationality that directs attention towards *transforming the expectations* of what actually *can* be delivered, given the water users' and the municipality's economic situation. Macleod explicitly talked about what he called the need to 'transform the thinking in communities':

> There is this culture in South Africa that is known as 'learned helplessness'. Historically, government told strictly to black people 'you have got no brains, you can't think, we will do everything for you'. They will tell you where you can live, they will tell you where you can go, they will tell you where you can work, they will decide what house you can live in.... 'Don't think if you think you are going to end up in jail. Mandela thought, look we've got him!' So they were thought to be helpless. And if the police came: your answer was 'I don't know' and you just acted dumb.
>
> (Interview with N. Macleod, 2009)

Macleod then explained that that way of thinking has continued in particular communities. The EWS, however, is working on changing such mentalities, he asserted, and gave the example that when people notice that taps are leaking they should report it to the council. He contrasted the mentalities in South African communities to other places around the world where community organizations are taking on tasks, organizing things for themselves. He expanded on this:

> if you think you are helpless and government must do everything, you really *disadvantage yourself*. If you say 'I am just me, I can't do anything'. *You and I don't think like that*, but millions of South African people do. And they are teaching their children that 'sit back, do nothing, act helpless, act dumb' and expect government, it's your right to expect the government [to do things for you]; provide you with grants to live, you see what it does to a society where they have these social grants, those people, some of them choose not to go and work: 'I can live, I can survive, on the grants that government gives me so why must I stress?' If we have a society of people like that, who would move the world forward? Who would solve the world's problems if nobody wants to work, earn a living, the hard way?
>
> (Ibid., my italics)

In Macleod's way of speaking about the qualities and capacities he thinks water users should have, there is a clear connection to the notion of 'self-reliance', which was critically discussed in Chapter 3 through the work of Duffield (2007) and Reid (2012, 2013). Macleod seems to align himself with the discourse of self-reliance that calls upon the water users' responsibility to change (and adapt to) the conditions of their lives. Here the (poor) subjects of water service delivery in South Africa appear as subjects that *make themselves powerless*, but who have the *potential* to be creative and take responsibility for their own lives. These capacities, however, need to be brought out. Notably, Macleod makes a clear distinction between himself (and me) in relation to such mentalities of helplessness.

Governing technologies: consultation, communication, and education

Hence, it is not only technological innovations that play a central role in the EWS management model, but also the 'social side' of water services delivery, as Teddy Gounden, Manager of Community Education, termed it.[11] The social side of water includes *consultation, communication, and education* with and for the water users. Such technologies of government can be seen as a way of transforming the water users, *through engaging them in their own governing*. These techniques are applied in order to change and transform the water users/citizens so that they can take responsibility for their own water consumption, and, through that process, arguably also for the municipality's water situation as a whole (cf. Cruikshank, 1999; see also Nash, 2013 for a Gramscian problematization of such governing).

There are two main rationalities behind working with such processes. First, it makes EWS' work 'easier'. Second, these practices are spoken of as an 'obligation', a logic that takes as its starting point notions of a rights-based service delivery whose target is to create water users who can hold the service providers accountable. Such a rationale is a paradigm shift – and therefore a massive challenge – compared to how water service delivery has been conducted previously, due to the legacy of colonialism and apartheid, which involved a 'top-down' approach.

EWS has teams who target communities where there are problems, for example, in terms of water loss. Through such practices, water users can be reached and informed about how they can be assisted, as well as what their responsibilities are. Additionally, as has already partly been indicated, pamphlets, road posts, and community radio are other ways of communicating water issues. Furthermore, information on water issues is provided via a toll-free number through which water users can, for example, report leaks and other faults as well as get information about their own water account.

Consultation is carried out through direct meetings in affected communities, and through water user forums and focus groups. Moreover, the idea of water conservation is a central pillar of what is termed 'consumer education', which is given when consumers get services installed, and is also done in a more formalized manner, through the water-training project.

The municipality's engagement in, so-called, water training was yet another way of working with perceptions. Called Raising the Citizens' Voice, this training is a collaborative effort between Mvula Trust (MT), the South African Water Caucus, DWAF, and the municipalities in which it has been active (SAWC and MT, 2009).[12] It aims to address public disengagement in water service delivery through a rights-based training module that develops into a dialogue between municipalities and their citizens (ibid.). In eThekwini the project was cofunded by the municipality, as a means of extending its existing public outreach programmes, and DWAF, to build capacity for civil society organizations.

The project has undertaken to educate councillors at all levels, civil society organizations, community committees, and the public. It covers the following themes:

Module 1 Introduction to basic legislation
Module 2 Different spheres of government
Module 3 Understanding the water cycle
Module 4 Life cycle of water from dams to taps and toilets
Module 5 Using water wisely
Module 6 Health, hygiene and sanitation
Module 7 Pollution abatement, and water quality
Module 8 Tariffs, billing and affordability
Module 9 Regulation, monitoring and evaluation
Module 10 Identifying gaps and planning the way forward

The municipality has taken an active part in the meetings by providing the participants with information and by taking part in discussions on problems related to service delivery. Gounden evaluated the project in the following words:

> now they [the participants] are really grasping the real issues, it's not just emotions but [they are] understanding the bigger picture, because we are governed by regulation. When they see the bigger picture they understand the challenges we are facing.

The project was also presented as a strategy to create a discussion where everyone (civil society, councillors, communities) would be able to start from a 'common understanding' of water service issues. This idea of creating a 'common understanding' is important for understanding the hydromentality of the EWS. The officials continually referred to the 'facts' and the 'reality' of service delivery when speaking of different (critical) viewpoints of EWS' service model. To them, a common understanding of what the reality is, is central to the task of communicating with water users and to the task of becoming a better and more efficient service provider.

Conclusion

This chapter has proposed the concept of hydromentality to make sense of different ways of defining water and how its proper governance and management is understood and practised.

The first section presented two major paradigms in water management. The first, the hydraulic mission, can be described in general terms as a hydromentality in which water was seen as a resource to build the nation state, its agriculture and industry, and to create modern citizens. Resource accessibility was, in many cases, taken for granted. However, in countries such as South Africa, water scarcity had become during the period of the hydraulic mission.

In this chapter the second paradigm has been termed neoliberal hydromentality. Such a conceptualization of present global water governance acknowledges the economizing aspects of water governance and places an emphasis on the interventions (the guaranteeing of basic human and ecosystem needs) that such a hydromentality employs to make it possible for market mechanisms to play a regulatory role, and on the different techniques that are made use of to govern both water use and water users. Thus, governing approaches that discursively lend themselves to inclusion and rights, and which view water as a 'public good', are to be understood as neoliberal in a governmentality sense.

At the centre of how both challenges and solutions to these challenges are understood, are the processes of drawing lines between different populations and between human and non-human life. Techniques that are promoted in order to govern water use and water users include participation and consultation, demand management and EBM, and technological devices. These techniques are employed so as to responsibilize water users, and to steer their behaviour and ways of understanding themselves.

The second and third sections of this chapter explored insights from governmentality studies of water. We learnt about: the distinct shapes that GWG has taken in particular places and contexts; how participation and inclusion in water governance have worked as techniques of governing; how subjectivities (both of decision-makers and water users) are created in hydropolitical relations of power; about the role that identities play in relation to water infrastructure and the access to water; and about the uncertainty of rule and of resistance.

The third section outlined the hydromentality of eThekwini. This particular hydromentality is characterized by the notion of water as a right and as a commodity, and by its practices of making distinctions between different water users through the applications of different technological artefacts and service levels for different individuals and populations. In this hydromentality, as is also indicated in some of the cases discussed in section two, we see how ways of understanding water that are different from those of the dominant understanding are 'othered' in the way that managers speak and that get their practical manifestations in awareness programmes and education interventions. The next chapter is entirely devoted to the biopolitical effects of water governance in this local context and how they can be studied through a combination of governmentality theory and narrative methodology.

Notes

1 Skilled labour was performed by whites who were paid more than other workers (black and coloured labourers) who performed the more unskilled tasks (Muller, n.d.).
2 The Orange River is the longest river in South Africa and plays an important role in the availability of water and the production of hydroelectricity. For a detailed description of the history of the river in general and of the Orange river project in particular, see Turton *et al.* 2004.
3 The fourth principle reads: 'Water has an economic value in all its competing uses and should be recognized as an economic good' (ICWE, 1992).
4 The concept of identity is here used in the sense of water's functions and roles.
5 Investment by the private sector in water services provision peaked in 1997 and then declined (Bakker, 2010, p. 92). In general, private involvement became more concentrated on smaller concessions, which did not require vast investment, such as management contracts (see Bakker, 2010; Kerf and Izaguirre, 2007). Investment was directed towards regions where overall investment was relatively high, such as Latin America and Asia, rather than low-income regions such as sub-Saharan Africa. This pattern also mirrors that of donor funding (Bakker, 2010, p. 94), leaving low-income regions underinvested. The reasons for this (albeit highly uneven) retreat of private involvement in the water sector include the relatively high capital investment that is required compared to other sectors, lack of effective regulatory capacity by different levels of government, and currency and revenue risks. Bakker notes that the 'political costs of raising tariffs to levels sufficient for cost recovery proved to be too great in a number of cases' (2010, p. 96). Moreover, it is acknowledged that protests against water privatization and commercialization had an impact on private investment (the most famous case being Cochabamba) in terms of raising costs and reducing interest in long-term concessions (ibid.).
6 The Bellagio principles read as follows:

> Recognize Water's Multiple Values Principle 1. Consider the multiple values to different stakeholders in all decisions affecting water. There are deep interconnections between human needs, economic well-being, spirituality and the viability of freshwater ecosystems that must be considered by all. Build Trust Principle 2. Conduct all processes to reconcile values in ways that are equitable, transparent, and inclusive of multiple values. Trade-offs will be inevitable, especially when water is scarce. Inaction may also have costs that involve steeper trade-offs. These processes need to be adaptive in the face of local and global changes. Protect the Sources Principle 3. Value and protect all sources of water, including watersheds, rivers, aquifers and associated ecosystems for current and future generations. There is growing scarcity of water. Protecting sources and controlling pollutants and other pressures are necessary for sustainable development. Educate to Empower Principle 4. Promote education and public awareness about the essential role of water and its intrinsic value. This will facilitate better-informed decision making and more sustainable water consumption patterns. Invest and Innovate Principle 5. Increase investment in institutions, infrastructure, information and innovation to realize the full potential and values of water. The complexity of the water challenges.
>
> (GWP, 2017)

7 The authors refer to an edition published in 1963.
8 This was the IDP in place at the time of the fieldwork.
9 This figure was changed later. In the 2016/2017 service level standards policy, properties with a value of more than ZAR250,000 are to be charged for the

usage of their first 9,000 litres of water (eThekwini Service level standards, 2016/2017).

10 During the fieldwork period, ZAR1 was equivalent to around US$0.1 (this was during a period of time when the value of the rand in relation to other currencies had declined steeply).

11 His full title (at the time of the fieldwork) was: Manager: Councillor Liaison and Community Education.

12 The Mvula Trust is a water supply and sanitation NGO; the South African Water Caucus is a civil society grouping in the water and sanitation sector; DWAF is the Department of Water and Forestry Affairs, which later became Department of Water Affairs (DWA) and Department of Water and Sanitation (DWS).

References

Abrahamsen, R. (2004) 'The Power of Partnerships in Global Governance', *Third World Quarterly*, 25(8), 1453–1467.

Agrawal, A. (2005) *Environmentality: Technologies of Government and the Making of Subjects*. Durham, NC: Duke University Press.

Allan, J. A. (2002) *The Middle East Water Question: Hydropolitics and the global economy*. London, UK: I. B. Tauris.

Alt, S. (2016) 'Beyond the Biopolitics of Development: Being, Politics and Worlds', PhD thesis, University of Lapland.

Ashton, P. and Haasbroek, B. (2002) 'Water Demand Management and Social Adaptive Capacity: A South African case', in Turton, A. and Henwood, R. (eds.) *Hydropolitics in the Developing World: A Southern African Perspective*. Pretoria, RSA: African Water Research Unit.

Babu, A. (2009) 'Governmentality, Active Citizenship and Marginalisation: The Case of Rural Drinking Water Supply in Kerala, India', *Asian Social Science*, 5(11), 89–98.

Bakker, K. (2004) *An Uncooperative Commodity*. Oxford, UK: Oxford University Press.

Bakker, K. (2005) 'Neoliberalizing Nature? Market environmentalism in water supply in England and Wales', *Annals of the Association of American Geographers*, 95(3), 542–565.

Bakker, K. (2007) 'The "Commons" Versus the "Commodity": Alter-globalization, Anti-privatization and the Human Right to Water in the Global South', *Antipode*, 39(3), 430–455.

Bakker, K. (2010) *Privatizing Water: Governance Failure and the World's Urban Water crisis*. Ithaca, NY: Cornell University Press.

Bakker, K. (2012) 'Water: Political, Biopolitical, Material', *Social Studies of Science*, 42(4), 616–623.

Barlow, M. and Clarke, T. (2003) *Blue Gold: The Fight to Stop the Corporate Theft of the World's Water*. New York, NY: Stoddart.

Birkenholtz, T. (2009) 'Groundwater Governmentality: hegemony and technologies of resistance in Rajasthan's (India) groundwater governance', *The Geographical Journal*, 175(3), 208–220.

Boelens, R., Hoogesteger, J., and Baud, M. (2015) 'Water reform governmentality in Ecuador: Neoliberalism, centralization, and the restraining of polycentric authority and community rule-making', *Geoforum*, 64, 281–291.

Bond, P. (2010) 'Water, health and the commodification debate', *Review of Radical Political. Economics* 42(4), 445–464.

Cremers, L., Ooijevaar, M., and Boelens, R. (2005) 'Institutional reform in the Andean irrigation sector: enabling policies for strengthening local rights and water management', *Natural Resources Forum*, 29(1), 37–50.

Crow-Miller, B., Webber, M., and Molle, F. (2017) 'The (re)turn to infrastructure for water management?', *Water Alternatives*, 10(2), 195–207.

Cruikshank, B. (1999) *The Will to Empower: Democratic Citizens and Other subjects*. Ithaca, NY: Cornell University Press.

Dean, M. (1999) *Governmentality: Power and Rule in Modern Society*. Los Angeles, CA: Sage.

Dean, M. (2010) *Governmentality. Power and Rule in Modern Society* (2nd edn). London: Sage.

Duffield, M. (2007) *Development, Security and Unending War: Governing the World of Peoples*. Cambridge: Polity.

Dugard, J. (2007) 'A pro-poor critique of prepayment water meters in South Africa: The Phiri story', *Critical Health Perspective*, 2007 nb 2.

EMWP (eThekwini municipality water policy). (2008/2009) [accessed 15 May 2009] No longer available online. In author's possession.

eThekwini Service level standards. (2016/2017) www.durban.gov.za/City_Services/water_sanitation/Service_Level_Standards/Documents/Service%20Level%20Standards%20Books%20English%20%20July%202016%202017%2010th%20Edition.pdf [accessed 17 January 2017].

EWS eThekwini Water and Sanitation (n.d.) *Most Innovative and Progressive water utility in Africa: Selected Highlights*. Booklet http://prg.ukzn.ac.za/docs/default-source/default-document-library/stockholm-water-week-brochure.pdf?sfvrsn=0 [accessed 27 October 2017]

Foucault, M. (2008) *The Birth of Biopolitics, Lectures at the Collège de France 1978–1979*. New York: Palgrave Macmillan.

Foucault, M. (2003) *'Society Must Be Defended', Lectures at the Collège de France 1975–1976*. New York, NY: Picador.

Freund, B. (2010) Is there such thing as a Post-apartheid City? *Urban Forum*, 21, 283–298.

Freund, B. and Padayachee, V. (eds.) (2002) *(D)urban Vortex: South African City in Transition*. Pietermaritzburg: University of Natal Press.

Funke, N., Nortje, K., Findlater, K., Burns M., Turton, A., Weaver, A., and Hattingh, H. (2007) 'Redressing Inequality: South Africa's New Water Policy', *Environment: Science and Policy for Sustainable Development*, 49(3), 10–23. DOI: 10.3200/ENVT.49.3.10-25.

Galvin, M. (2016) 'Leaving boxes behind: civil society and water and sanitation struggles in Durban, South Africa', *Transformation: Critical Perspectives on Southern Africa* 92, 111–134.

Gandy, M. (2004) 'Rethinking urban metabolism: Water, space and the modern city', *City*, 8 (3), 363–379.

Gandy, M. (2006) 'Zones of Indistinction: Some thoughts on the bio-politics of urban space', *Cultural Geographies*, 13(4), 1–20.

Gandy, M. (2014) *The Fabric of Space: Water, Modernity, and the Urban Imagination*. Cambridge, MA: The MIT Press.

Gounden, T., Pfaff, B., Macleod, N., and Buckley, C. (2006) *Provision of Free Sustainable Basic Sanitation: The Durban Experience*. Sri Lanka: 32nd WEDC International Conference, Colombo.

GWP (Global Water Partnership). (2017) High Level Panel on Water Valuing Water Initiative www.gwp.org/en/we-act/campaigns/high-level-panel-on-water-valuing-water-initiative/, [accessed 5 October 2017].

Hansson, S. (2013) 'Who brings the water? Negotiating state responsibility in water sector reform in Niger', PhD Thesis, University of Gothenburg.

Hansson, S. (2014) 'Dependence and Autonomy: Conditions of Possibility for Nigerien Ownership in the Water Sector', *Journal of Intervention and Statebuilding*, 9(2), 256–277.

Hansson, S., Hellberg, S., and Stern, M. (eds.) (2015) *Studying the Agency of Being Governed: Methodological reflections*. London, UK: Routledge.

Hellberg, S. (2014) 'Water, life and politics: Exploring the contested case of eThekwini municipality through a governmentality lens', *Geoforum*, 56, 226–326.

Hellberg, S. (2015) 'The Biopolitics of Water: Technology, Subjectivity and Lifestyle in eThekwini municipality', PhD Thesis, University of Gothenburg.

Hellberg, S. (2017) 'Water for Survival, Water for Pleasure: A biopolitical perspective on the social sustainability of the basic water agenda', *Water Alternatives*, 10(1), 65–80.

Hey, J. and Klak, T. (1999) 'From Protectionism Towards Neoliberalism: Ecuador across four administrations (1981–1996)', *Studies in Comparative International Development*, 34(3), 66–97.

Hoogesteger, J. (2012) 'Democratizing Water Governance from the Grassroots: The development of Interjuntas-Chimborazo in the Ecuadorian Andes', *Human Organization*, 71(1), 76–86.

ICWE (International Conference on Water and the Environment). (1992) *The Dublin Statement on Water and Sustainable Development*. Dublin, IE: ICWE.

IDP (Integrated Development Plan) (2006/2007) eThekwini Municipality Integrated Development Plan 2006/07 www.durban.gov.za/Documents/City_Government/IDP_Policy/IDP_2006_07.pdf [accessed 16 January 2017].

IDP (2016/2017) eThekwini Municipality Integrated Development Plan 2016/2017 www.durban.gov.za/city_government/city_vision/idp/documents/final%20 2016_17%20IDP%2029052016.pdf [accessed 16 January 2017].

IWA (The International Water Association) (2013) Newsletter, September.

Kaika, M. (2005) *City of Flows: Modernity, Nature and the City*. London, UK: Routledge.

Kerf, M. and Izaguirre, A. (2007) 'Revival of Private Participation in Developing Country Infrastructure. A Look at Recent Trends and Policy Implications', *Gridlines 16*. Washington DC: World Bank, Public Private Infrastructure Advisory Facility.

Kooy, M. and Bakker, K. (2008) 'Technologies of Government: Constituting Subjectivities, Spaces, and Infrastructures in Colonial and Contemporary Jakarta', *International Journal of Urban and Regional Research*, 32(2), 375–391.

Latour, B. (1993) *We Have Never Been Modern*. Cambridge, MA: Harvard University Press.

Legg, S. (2006) 'Governmentality, congestion, and calculation in colonial Delhi', *Social and Cultural Geography*, 7, 709–728.

Linton, J. (2010) *What is Water? The History of a Modern Abstraction*. Vancouver, BC: UBS Press.

Loftus, A. (2005a) 'A political ecology of water struggles in Durban, South Africa', PhD thesis, University of Oxford, UK.

Loftus, A. (2005b) '"Free Water" as commodity: The paradoxes of Durban's water service transformations', in McDonald, D. A. and Ruiters, G. (eds.), *The Age of Commodity: Water Privatization in Southern Africa*, pp. 189–203. London and Sterling: Earthscan.

Loftus, A. (2006) 'The metabolic processes of capital accumulation in Durban's waterscape', in Heynen, N., Kaika, M., and Swyngedouw, E. (eds), *In the Nature of Cities: Urban Political Ecology and the Politics of Urban Metabolism*, pp. 173–190. London and New York: Routledge.

Loftus, A. (2007) 'Working the socio-natural relations of the urban waterscape in South Africa', *International Journal and Regional Research* 31(1), 41–59.

Loftus, A. (2009) 'Rethinking political ecologies of water', *Third World Quarterly* 30(5), 953–968.

Mayhew, S. (ed.) (2004) 'Governmentality', in *A Dictionary of Geography*. Oxford, UK: Oxford University Press.

Mirosa, O. and Harris, L. (2012) 'Human Right to Water: contemporary challenges and contours of a global debate', *Antipode*, 44(3), 932–949.

Molle, F., Mollinga P. P., and Wester, P. (2009) 'Hydraulic Bureaucracies and the Hydraulic Mission: Flows of Water, Flows of Power', *Water Alternatives*, 2(3), 328–349.

Mirumachi, N. and van Wyk, E. (2010) 'Cooperation at Different Scales: Challenges for local and international water resource governance in South Africa', *The Geographical Journal*, 176(1), 25–38.

Morinville, C. and Rodina L. (2013) 'Rethinking the Human Right to Water: Water access and dispossession in Botswana's Central Kalahari Game Reserve', *Geoforum*, 49, 150–159.

Muller, M. (n.d.) *Lessons from South Africa on the management and development of water resources for inclusive and sustainable growth*. European Report on Development.

Narsiah, S. (2010) 'The neoliberalisation of the local state in Durban, South Africa', *Antipode* 42(2), 374–403.

Nash, F. (2013) 'Participation and Passive Revolution: The reproduction of neoliberal water governance mechanisms in Durban, South Africa', *Antipode* 45(1), 101–120.

Ong, A. (2007) 'Neoliberalism as a mobile technology', *Transactions*, 32, 1–6.

Ospina-Peralta, P. (2009) *Transiciones. Ecuador después del referendum*. Comité Ecuménico de Proyectos, Quito.

Pollard, S. and du Toit, D. (2011) 'Towards Adaptive Integrated Water Resources Management in Southern Africa: the role of self organisation and multi scale feedbacks for learning and responsiveness in the Leteba and Crocodile catchments', *Water Resources Management*, 25, 4019–4035.

Reid, J. (2012) 'The disastrous and politically debased subject of resilience', *Development Dialogue,* 58, April.

Reid, J. (2013) 'Interrogating the Neoliberal Biopolitics of the Sustainable Development-Resilience Nexus', *International Political Sociology*, 7(4), 353–367.

Rogers, S., Barnett, J., Webber, M., Finlayson, B., and Wang, M. (2016) 'Governmentality and the Conduct of Water: China's South–North Water Transfer Project', *Transaction of the Institute of British Geographers*. Doi: 10.1111/tran.12141

Rutherford, S. (2007) 'Green governmentality: insights and opportunities in the study of nature's rule', *Progress in Human Geography*, 31(3): 291–307.

RSA (Republic of South Africa). (1962) *Report on the Proposed Orange River Development Project*. Pretoria, RSA: Department of Water Affairs. Doc. No. W.P. X'62.

RSA. (1970) *Report of the Commission of Enquiry into Water Matters*. Pretoria: Government Printers. Doc. No. R.P. 34/1970.

RSA. (1996) *Constitution of the Republic of South Africa*, Act 108 of 1996. Pretoria: Government Printer.

RSA. (1997) *Water Services Act*, no. 108 of 1997. Pretoria: Government Gazette.

RSA. (1998) *National Water Act*, no. 36 of 1998. Pretoria: Government Gazette.

RSA. (2001) Gazette No. 22355, Notice R509 of 2001 (8 June 2001) published in terms of section 9 of the South African Water Services Act 108 of 1997. Pretoria: Government Gazette.

SAWC (The South African Water Caucus) and MT (Mvula Trust). (2009) *Civil Society Involvement in the Citizen's Voice in Water Service Regulation in the eThekwini Municipality (Case Study)*. Final Report.

Shiva, V. (2002) *Water Wars: Privatization, Pollution and Profit*. Cambridge, MA: Sound End Press.

SIWI (Swedish International Water Institute) (2014) 'Press Release "Most progressive water utility in Africa" wins 2014 Stockholm Industry Water Award' www.siwi.org/prizes/stockholmindustrywateraward/winners/2014-2/, (accessed 30 May 2014).

Sultana, F. and Loftus, A. (2012) *The Right to Water: Politics, Governance and Social Struggles*. London, UK: Routledge.

Sutherland, C., Scott, D., and Hordijk, M. (2015) 'Urban Water Governance for More Inclusive Development: A reflection on the 'waterscapes' of Durban, South Africa', *The European Journal of Development Research* 27(4), 488–504.

Turton, A. R., Meissner, R., Mampane, P. M., and Seremo, O. (2004) *A Hydropolitical History of South Africa's International Rivers Basins*. African Water Issues Research Unit (AWIRU) Pretoria, RSA: University of Pretoria.

Ward, L. (2013) 'Eco-governmentality Revisited: Mapping divergent subjectivities among Integrated Water Resources Management experts in Paraguay', *Geoforum*, 46, 91–102.

Wittfogel, K. (1957) *Oriental Despotism; a Comparative Study of Total Power*. New York, NY: Random House.

WSDP (2004) *Water Services Development Plan*, eThekwini Water and Sanitation Unit, Vol. 2. No longer available online. In author's possession.

WRC South Africa (Water Research Commission) (2011) 'An international award for Ethekwini's water conservation and sanitation', www.wrc.org.za/News/Pages/AninternationalawardforEThekwini'swaterconservationandsanitation.aspx [accessed 27 October 2017].

WWDR (World Water Development Report). (2015) *Water for a Sustainable World*. UNESCO, http://unesdoc.unesco.org/images/0023/002318/231823E.pdf [accessed 8 October 2017].

WWDR (World Water Development Report). (2016) *Water and Jobs*. UNESCO, http://unesdoc.unesco.org/images/0024/002439/243938e.pdf [accessed 8 October 2017].

WWDR (World Water Development Report). (2017) *Wastewater the Untapped Resource*. UNESCO, http://unesdoc.unesco.org/images/0024/002471/247153e.pdf [accessed 8 October 2017].

ZACC (Constitutional Court of South Africa). (2009) CCT 39/09 [2009] ZACC 28: *Mazibuko and Others* v. *City of Johannesburg and Others*, www.saflii.org/za/cases/ZACC/2009/28.html [accessed 11 October 2017].

List of interviews

Gounden, T. (2008) Manager: Councillor Liaison and Community Education, Water and Sanitation, eThekwini municipality. Interviewed on 17 and 22 October.

Macleod, N. (2009) Head of Water and Sanitation, eThekwini municipality. Interviewed on 2 April.

Singh, M. (2008) Deputy Head, customer services at eThekwini Water and Sanitation. Interviewed on 11 December.

5 Understanding biopolitical 'effects' locally

Life(styles), subjectivities, and moral obligations

Introduction

The previous chapter outlined the hydromentality of eThekwini municipality as characterized by the different technologies that have been developed and applied to suit different areas, populations, and individuals: ground tanks have been installed in rural areas where infrastructure is too costly to invest in; flow limiters in areas where people have accumulated debts (and where 'illegal connections' have been widespread); standpipes in informal settlements; and regular meters in the suburbs and townships. In biopolitical terms, such targeting of certain types of population as suited to particular technological solutions can be understood as a way of distinguishing between different forms of life. Furthermore, the chapter emphasised how the municipality employs techniques that engage the water users themselves in their own governing, including consultation, communication, and education.

This chapter explores these techniques of governing in terms of their biopolitical effects, outlines a methodological framework for studying the effects of a particular hydromentality, and offers a close reading of people's understanding of themselves, their lifestyles, their place in society, and their responsibilities as humans/individuals/citizens. Through such a local study, we learn about the complex effects of water governance in terms of what water and the way it is delivered mean for people's lives and lifestyles, as well as how notions of sustainability and responsibility are produced.

The first section presents the methodological framework that was applied in studying the biopolitical 'effects' of the hydromentality of eThekwini, its use of narrative methods, and how a biopolitical reading of the narratives was conducted.

The second section outlines what we can learn from a biopolitical analysis of water narratives. It begins by stressing two basic points regarding the emphasis that the methodological framework places on, first, the uncertainty of governing and the agency of the water users, and second, the importance of a *historicized* understanding of water and water infrastructure.

The second section is divided into four parts. The first part discusses productive effects of water and water technologies in terms of lifestyles and subjectivities. The second part outlines how the narrators, through their stories, make sense of the right to water and the way that FBW policy is implemented. The third part discusses the making of (un)sustainable and (ir)responsible water users in hydropolitical relations of power, while the fourth part engages in narratives on the water training project. This is followed by some concluding remarks.

Studying biopolitical 'effects'

When concerned with the subject of governance, governmentality and biopolitics studies often remain focused on how such strategies of governing target certain populations and individuals. In this book, investigating the biopolitics of water not only includes studying the hydromentality, it also involves looking into the effect of such governing logics in terms of water's role in people's lives and how water users engage with hydropolitical strategies of governing. Since this is methodologically challenging the chapter presents a framework for studying biopolitical effects of water management (for more reflections on this methodology see Hellberg, 2015).

Narrative methodology

In order to study what water and water governance mean in terms of people's lives, a methodological approach that allows for getting close/r to the water users' experiences and perceptions is needed. This book therefore makes use of narrative methodology.

Such an approach allows for a focus on the water users' experiences, *as they expressed them in their stories* (cf. Hinchman and Hinchman, 1997, p. xvi). This methodology offers a way of creating texts about how water users make sense of their (water) situation and relay a sense of self (Elliott, 2005, p. 126; Hinchman and Hinchman, 1997, p. xvi). However, stories are not mirrors of reality, as recognized by narrativists (Hinchman and Hinchman, 1997, p. xvi). Rather than representing 'real life' (White, 1987, p. ix), they are constructive of reality. They simplify, rearrange, and redescribe elements and events in order to mediate between the self and the world and to create order and meaning (Hinchman and Hinchman, 1997, p. xvi). Hinchman and Hinchman offer the following definition:

> Narratives (stories) in the human sciences should be defined provisionally as discourses with a clear sequential order that connect events in a meaningful way for a definite audience and thus offer insights about the world and/or people's experiences of it.
>
> (Ibid.)

According to the above definition, the meaning of a narrative becomes constructed through the interaction between the narrator and her or his audience and can be seen as a product of negotiations between the listener and the narrator (ibid., 9). Narratives are also constructed, however, in relation to the narrator him- or herself, as the narratives can be understood as a way of creating a 'sense of self' (Elliott, 2005, p. 126).

Another basic assumption of narrative inquiry that we also understand through the definition above, is the close connection between narrativity and temporality (Johansson, 2005, p. 84). Through our narratives we place ourselves in time (ibid., based on Ricoeur, 1991). Or in Hinchman and Hinchman's words:

> we continuously create and reinforce our sense of self by linking our present plans, actions, and states to both the future (as 'project') and the past, as the already articulated story of our lives to that point.
>
> (Hinchman and Hinchman, 1997. p. xviii)

Yet another important point to stress as regards how we are to assess narratives in relation to governing strategies, is that the narrator's voice is to be seen as a 'voice of authority' in relation to the reader or listener (Carr, 1997, p. 13). This means that even though a narrative is a product of negotiation, the narrator should be regarded as the 'author of his or her own story' and 'an active shaper of outcomes' (Hinchman and Hinchman, 1997, p. xix).

According to this approach, agency is stressed and the self is seen as active and self-shaping, which, in turn, opens up possibilities for analysing how water users actively and productively engage with relations of power in water service delivery. Here, these narratives are treated as instances where the self takes shape in relation to different discourses and power relations (including that of the relation between interviewer and interviewee).

Studying processes of subjectivation (which can be seen as processes within which subjects turn their gaze on themselves) and the governing of the self, there is, however, a need to problematize the possibility of pinning down agency as well as the 'effects' of power. In a framework where power is seen as '*forming* the subject as well as providing the very condition of its existence' (Butler, 1997, p. 2, italics in original), it is not possible to distinguish between 'the power that forms the subject and the subject's "own" power' (ibid., p. 15), as an 'irresolvable ambiguity' (ibid.) arises. Of relevance here are the two meanings of the subject according to a Foucauldian understanding. In 'The Subject and Power' Foucault writes:

> There are two meanings to the word 'subject': subject to someone else by control and dependence, and tied to his own identity by a conscience or self-knowledge. Both meanings suggest a form of power that subjugates and makes subject to.
>
> (2000, p. 331)

Consequently, the subject is at the same time the target of power and an actor whose conditions of existence are provided by power (Butler, 1997, p. 2).

Hence, within a framework that acknowledges this 'ambivalent sense of agency' (ibid., p. 15), a number of difficult methodological issues arise with respect to analysing agency. For example, how are we to understand agency when water users act in ways that correspond to governing logics?

According to the framework developed in this book, the possibility of thinking and acting differently in relation to governing logics has been acknowledged as a form of agency, but so has the active work that people do on themselves in relation to different techniques of power. Therefore, we can approach the water users' stories – whether they resist governing rationales or not – as stories through which we can study the 'agency of being governed' (see Hansson, *et al.*, 2015). What this methodology successfully brings into focus is the interconnectivity between government and self-construction. In line with the above, we should nevertheless be wary of the fact that what an interviewer gets from people's stories is far from a representation of how (biopolitical) power relations play out in their everyday lives in their totality. Rather, these stories can be understood as narratives in which water users attempt to act and position themselves in relation to technologies of power, and to a particular audience, *in the moment of telling their stories.*

Acknowledging this means that what the narrators chose to tell me was not only based on what they were willing to share with a stranger, but it was also dependent on what they thought was meaningful for me to hear, hence, what they thought I *wanted* and *could* hear. One man, Mr Mhaule, explicitly addressed this, asking 'where it would take her [me] to know' about past water practices. His initial objection to telling such stories can also be understood in relation to modernity in general and the dominance of 'modern water' in particular (Linton, 2010), as 'traditional' water practices become viewed as both irrelevant and irrational in relation to modernity. Mr Mhaule said:

> things like that are not easy to tell and make into a story … because these things vanish. I don't know … it is not a story you can tell … because it is something that ends, nothing [that will] happen again.

Interviews with 64 narrators were conducted in four different areas of the municipality: one rural community (Mzinyathi); one peri-urban/township area (Inanda/Ntuzuma); one 'urban' township (specifically the communities of Westcliffe and Crossmoor in Chatsworth); and suburban middle-class areas (Berea, Glenwood, Westville, Merbank Ridge, and Wentworth Assegai). The areas were selected after conducting an inventory of different service levels and systems. Both areas and narrators in these communities were chosen to include: as many different ways of accessing water as possible; a diversity of informants regarding ethnicity, age, gender, household size; and a geographical spread in the community. In the following text, pseudonyms are used.

Having learnt from narrative methods, interview questions were organized around temporal dimensions; asking how water service delivery and the technologies and artefacts through which they received water had changed during the informants' lives; encouraging them to weave their life stories together with issues relating to water and to tell both their water stories/water memories and their associations with water and 'water places' (cf. Gibbs, 2009), such as dams, rivers, lakes, and streams. They were specifically probed to tell stories about technological artefacts (tanks, meters, standpipes) and tariffs. They were asked what they saw and wished for themselves in the future in terms of water, or if they wanted anything to be different. This way of getting narrators to talk about future fears, hopes, and wishes proved to be a good way of probing how they thought about their present situation. The interviews were conducted together with a field assistant, who also, when needed, acted as a translator from isiZulu into English (in Mzinyathi and in Inanda/Ntuzuma). These interviews were carried out at the end of 2008 and in 2009. The following quotations have been edited for readability.

A biopolitical reading of water narratives

In order to conduct a systematic biopolitical reading of the narratives, a set of questions were posed to the interview material. *In the water users' narratives*, how are water and water service delivery/technologies of water management related to:

> **the notion of life:** How are the narrators telling stories of life, death, survival? In what ways do the narrators relate water to (the life of) nature/the environment/animals/humans? How are the narrators relating water to the body/biology? To health/sickness/cleanliness/pollution? To fears/worries? (How) is life made political in relation to water? How do notions of rights feature in relation to water/life? How do the narrators relate to the idea of payment for water and the right to water?
>
> **ways of life/lifestyles:** How are the narrators talking about their and others' ways of life? In what ways do the narrators articulate that water service delivery is a limiting/enabling factor in their everyday lives? How do they relate water to modernity? To urban/rural ways of life? To the meaning of places and locations? To poverty/wealth? To pleasure and recreation? To guilt? To (un)dignified life? To well-being and the good/miserable life? To responsible/sustainable ways of lives? How does the theme of materiality feature in their stories?

identity, subjectivity, responsibility, and agency: How do the narrators articulate resistance, experiences of exclusion/inclusion, ideas relating to inside/outside? To notions of class/race/gender? To citizenship? How do they relate water to notions of community/the state/the individual? In what ways do the narrators articulate certain ideas, logics, and values in relation to water and technologies of water service delivery? To whom (or what) do people ascribe agency? In what ways do the narrators construct themselves as active shapers of outcomes or as powerless? How do the narrators produce images of what kind of water user they are and/or would like to be? How do they relate to notions of responsibility and sustainability? Do they (and how) relate to scarcity/waste of water?

Exploring the biopolitics of water through (bio)narratives

Below is an outline of what we learnt from a biopolitical analysis of the water narratives, starting with two basic points related to, first, the agency of the water users, and, second, the importance of a historicized understanding of water and water infrastructure.

A first basic point regarding what we learn from the water stories through the chosen methodological approach, is how difficult it is to conduct a tidy analysis of power and the subjectivities of governing strategies because of the messiness and complexity of empirical data. Nonetheless, this should not be seen as an analytical weakness. Rather, such an analysis places a focus on the uncertainty of governing and the different ways in which people engage with assemblages of hydropolitical power relationships. The water users in eThekwini are not just passive subjects who are successfully governed to behave in certain ways; they are agents who actively navigate their lives.

Thus, the framework calls for attention to the agency of the water users in hydropolitical relations of power, where the materiality of both water and technological devices play active parts. It is a way of attuning to difference, to unintended consequences, and to acts of resistance (Rutherford, 2007, p. 305) in relation to the particular hydromentality of eThekwini. It is also used as a tool for problematizing how the water users' agency is utilized in their own governing and how these strategies of governing play a part in producing particular lifestyles, subjectivities, and notions of sustainability and responsibility, which is discussed below.

Another basic starting point for an understanding of how water users understand their present (lack of) water services is that the past is always inscribed in narratives of present and future water situations (see also Strang, 2004). In eThekwini, personal water life (hi)stories reflect experiences of discrimination and of privilege. This means that when assessing the water users' understandings of, and claims to, water, the past has to be reckoned with. Such understandings have markers in the landscape. Bakker expressed this by:

> 'things' (pumps, dams, canals) ... make a difference for the way social relations unfold, [they] are not merely pre-given substrates that enable and constrain social action; rather, they are themselves historically and geographically produced in a way that is simultaneously socio-natural and socio-technical.
>
> (Bakker, 2012, p. 621)

Through exploring people's narratives of water infrastructure, such as dams, we can understand, in the South African context, not only how colonialism, racism, and nationalism are inscribed in landscapes but also how such 'water places' can be understood as symbols of modernity and progress. Associations with the past are carried into the ways people inscribe present and future images of water service delivery and make sense of their lives as a whole in their recollections. In turn, such insights can provide perspectives on the legitimacy of a water governance that differentiates between different populations and users on the basis of conditions that were shaped during the apartheid era.

One narrator, O'Brien, told a story characterized by feelings of loss, both of livelihood and community relationships, in relation to the building of the Inanda dam.[1] When asked what he thought of when he saw the dam, he responded:

> I feel like crying because you know even today as I am here [in Inanda] as I dream, I don't dream here. I still dream there.... Every time I go past the Inanda dam, you know, I think about our relatives' bones and bodies. I think about friends that we had. I think about respect that we used to have. I think about schools. I think about churches. I think about the friendship that was actually broken because now: if I want to go to KwaNgcolosi ... I need transport, I cannot just go, but I used to ... we used to visit there. If they had got a function, they used to call us, we just walked across the river ... and came back. When they were building the Inanda dam they did not think that we have got relatives that we would like to visit, so all of that friendship is broken because now either I go via Durban, then Pinetown, then Newlands, then KwaNgcolosi.... It's a long way, so these relationships have been broken because of the dam.

The dam was placed in between O'Brien's home and the neighbouring KwaNgcolosi, which cut off easy access between the communities.[2] O'Brien tells a horrific story about witnessing workers digging up bodies from a graveyard to make a place for the dam; hence O'Brien's associations to relatives' 'bones and bodies'.

Through this account, O'Brien constructs his story around memories in which the 'white man' symbolizes repression and the dam is a symbol of death, both literally, as in the particular story above, and symbolically, in terms of the friendships and relationships that have been lost. The dam is hence an icon for the disrupted relationships within the community and between neighbouring villages, for the lost connection and feeling of belonging to the land, and for the lost relationship to the dead.[3]

To other narrators, memories of dam construction had other connotations, however. Dams were to be admired. One narrator, Zoe, living in the suburb of Berea, told me about her particular relationship to dams and dam construction. A central part of Zoe's water narrative is her upbringing. As a child of a water engineer, she allegedly grew up on different dam sites. The family travelled from one construction site to another. Zoe said:

> You know there were days where we would always go pass the dam and stop at the dam, walk around the dam *[laughing]*. All the family photographs are of construction sites, you know.... Which is typical if you have a parent who is an engineer.

Such memories are, as Zoe put it, from 'a period in our history where the construction of dams was seen as quite a rational thing to do'. In her narrative, she showed great awareness of the different phases of water governance that South Africa has gone through. Her statement is symptomatic of the rationality that prevailed during the, so-called, hydraulic mission, which was immensely important in the construction of the South African state during the twentieth century. For white families such as Zoe's, visits to the dam can be understood as a way of celebrating modernity and progress, as these processes took shape for the privileged during the apartheid era. This initial discussion of the different relationships to dam construction has the purpose of illustrating the importance of an historicized understanding of water and water infrastructure for appreciating how people relate to water service delivery in the present.

Productive effects of water technologies: lifestyles and subjectivities

Through examining people's water stories, it is evident that the relationship between water and the notion of 'life' goes beyond water's role in maintaining survival. While the physical need for water to survive lurks in the background of most narratives, more prominent themes are how water makes people feel about themselves (for instance, respectable, non-human, privileged, excluded) and their life situation (for example, convenient, a struggle, modern). In this sense, 'water isn't just water' (Gibbs, 2006, cited in Linton, 2010) and its relationship to life extends far beyond sustaining life (*zoë*), and is fundamental to the construction of the proper or the good life (*bios*) (cf. Agamben, 1998). In water users' narratives, water both represents and actively shapes (notions of) the self (and others), as well as lifestyles. It is productive of the very lifestyles that people lead (or have led) and of ideas surrounding other people's lives and lifestyles. This is discussed in the text below, which also emphasises how the materiality of water, and hydropolitical relations of power, are coconstitutive of identities and subjectivities (see Sultana, 2009), a theme we return to in the next section on 'making sense of the right to water'.

Stories of rural/urban lives

The relationship between water and lifestyle, and its role in creating subjectivities, is evident in Dumisani's narrative. Dumisani grew up in a rural area and at the time of the interview was living in a flat in the suburb of Glenwood, enjoying the comfort of modern water facilities. He talked about the difference in water access in terms of living two different lives:

> Water is life ... water sustains life. It's a very basic necessity for everyone ... I feel sorry for communities that do not have access to water ... how do they keep life with limited access to water? ... Do they do things that people ... normally do with water if they don't have access to it? Life ... has not changed for those and I feel sorry for them. *I think having lived two lives*, a life in rural area where my home is, and life here in the urban area and I know that it's not the same ... there are communities that don't have access to water and I do not know how do they live.... With life, water is everything.[4] (My italics)

Several points relevant to a biopolitical analysis can be extracted from this excerpt. In Dumisani's story, water is *constitutive of* the lifestyles. The excerpt also shows how we can understand processes of the *normalization* of urban ways of accessing and using water in everyday life; even though Dumisani had grown up in a rural area, he narrated these practices as unimaginable ('I do not know how they live', he said). Lastly and relatedly, through this story Dumisani constructed himself as an urban and modern subject whose life had transitioned from a rural past. Such stories of the rural–urban difference in terms of water access and use constitute a main storyline in many of the narratives, as do the themes of modernity and proper ways of living in the present.

One narrative (albeit told in a different way) centred on the hard work that was involved in carrying water long distances from the river. At the same time, however, the narrators spoke enthusiastically about the social life and the cultural practices taking place by the rivers. The nostalgia that surrounded the way some of the water users told stories of water in the past should not be interpreted as meaning that they would prefer to still rely on the river as a source of water. For many of the narrators – not only in rural Mzinyathi, but also in other sites – the remembering of past water practices served a purpose in upholding collective memories and identities. These identities are, however, not static, as pointed out by Hall (2008);

rather, they can be understood as 'fluid' (Stern, 2005, 2006). What is understood as appropriate in the past is not necessarily seen as suitable in the present, as most narrators embraced the need to be provided with piped water from a utility, a theme to which we will return.

Stories of struggle and of 'zones of (in)distinction'

In the townships, the inhabitants relied on different service levels and ways of accessing water (conventional water meters, 'illegal' connections, standpipes, water vendors, and flow limiters [for an explanation of the different technologies and service levels, see Chapter 4; all of these ways of accessing water will be referred to throughout the chapter]).

Those who had a conventional water meter often related water access to payment. For many, this involved a struggle to make ends meet. Mrs Bongani, living in Ntuzuma, explicitly addressed the difficulty of changing from freely available water in rivers and streams to a system that requires payment:

> You see there in *the changing of life* when you are paying for water when you were not paying for it, it is hard.... We are old now, you are now supposed to sort it out. *We grew up in the rural areas, not doing all these things of paying for water* ... now they are robbing us, it is robbery what they are doing when you [have to] pay for electricity and pay for water. (My italics)

For others, who relied on either water vendors or illegal connections, access to water represented a daily struggle, even though their difficulties in relation to water were different. Mrs Ngema, who lived in a self-made shelter in Inanda White City, and who had been waiting for over two years for an RDP house (a cheap government house in the reconstruction and development programme), explained that her family paid ZAR 10 to the neighbours in order to access water from them.[5] The neighbours' connection, in turn, was (illegally) connected to the standpipe on the road. The problem, Mrs Ngema told us, was that the neighbours regulated access and that Mrs Ngema therefore could not access water until the neighbours had woken up in the morning.

In an informal settlement in Crossmoor, struggle is a prevalent theme. A standpipe had been installed to provide the inhabitants with their FBW. Before its installation, people in the settlement relied on access from private houses situated close to the settlement. The instalment of the tank improved conditions in the settlement considerably. Nevertheless, the inconveniences and difficulties of the life people led in the community were common themes in the narratives from this particular site. Samuel, an elderly man, said: 'being here, you sometimes feel like you are an animal. You can't freshen up.' Continuing to speak of the past, when he had access to a shower and a flushing toilet, he said with emphasis: 'that's a way of living!'

Samuel's remarks relate directly to the biopolitical problematique regarding the distinction between what is regarded as a (dignified and proper) human life and living like 'an animal'. Samuel narrated himself as a figure who has been granted formal citizenship and incorporated into the system of rights, and yet has been stripped of humanity.

Furthermore and relatedly, conditions in the settlement were narrated as depriving people of their individuality. For example, Ameena, a young Muslim woman, explicitly talked about how the material conditions in the settlement prevented her from living according to the prescriptions of her faith. She explained that she used to cover her face, which no longer made sense as people lived so closely together in the settlement, and the difficulties of keeping to the rules regarding different uses of water and praying.

Informal settlements, such as the one in Crossmoor, are probably the places in which the paradoxes of South African (political) life can be explored to the fullest. They can be understood as 'zone[s] of irreducible indistinction' (cf. Agamben, 1998, p. 9), where bare life is produced through inclusion/exclusion. What it means in practical terms, in Crossmoor, is that official inclusion in a political system of rights (to water/life) coincides with the exceptional state of being and living as part of a population without the possibility for individual and private lifestyles or preferences. Hence, at the same time as being captured by the political order, the people in the settlement are excluded from political life/proper life/good life (Agamben, 1998; Plonowska Ziarek, 2012).

Stories of privilege

The contrast between material conditions in an informal settlement and those in suburban areas is striking, even provoking. While narrators from the informal settlement tell stories that express how material conditions, such as the lack of water facilities and private space, prevent them from feeling human, interviewees from the suburbs narrate water as something essential in terms of their privileged lifestyles, and in terms of who they are as middle-class South Africans. Karen, living in Glenwood, for example, narrated water as intimately connected to her lifestyle, which she described as 'elite', in which holidays and social life revolved around water: beaches, swimming pools, rivers, and waterfalls (as indicated in the introductory pages of the previous chapter). She said:

> [M]y whole life has been about water … I can see that for my children too. If you watch the videos that we were watching this morning from when they were one year old, you see them first going to the beach on the north coast and then, we had this holiday on the sea. Then you see

> us going to the beach on the south coast. And there is them swimming in the sea. Then you see us going to the Drakensberg and we take them to the rivers and they swim in the rivers and under the waterfall ... and then we come back to the camp and it's hot and so we put them in buckets with water and, and all the video footages of them in water and it's incredible ... and in the swimming pool here at our house, in the bath upstairs. There is an association for me between happiness and water ... and positive experiences and water. I mean of course they are all leisure things, and are all the kinds of things people can do when they have leisure time and leisure money and you know, I recognize that, but that has certainly been my experience as a white South African who grew up under a privileged system of apartheid where whites just had access to holidays and money and that sort of thing, *I mean everything we ever did was about water.* (My italics)

In many of these narratives, examples were taken from contexts understood to be radically different from one's own, in order to illustrate aspects of what water use meant for their lifestyles. The narratives thus oscillated between the everydayness of one's own water use and the extraordinariness of that of others. Hence, a certain kind of exotification was common in many of these stories and in turn produced subjectivities (in the suburbs) such as the privileged, modern, or urban water user. In biopolitical terms, such narratives can be understood as a way for these water users to display the difference between *zoë* and *bios*, that is, the difference between 'mere' life and a 'proper' or 'good' life (Agamben, 1998; Plonowska Ziarek, 2012, p. 1) and, furthermore, to produce the image of themselves as belonging to a 'qualified' form of life (Wrangel, 2014, p. 187). Lauren illustrated the differences between different ways of living in relation to water access:

> I consider myself to be very lucky because we have things like a roof over our head and running water and electricity, like, you know, the basics, we have the basics ... [I don't have] to think about it. And I think it is important to teach the children about that as well but that isn't just something that should be taken for granted.

Ideas surrounding water and modernity and a modern lifestyle are features in both Lauren's and other narrators' stories. These stories are also gendered. For Lauren, who said that her family 'never [has] to worry about water', water and water services are related to ideas of modernity. When asked whether identity had implications for her relationship to and usage of water, Lauren, who said she saw herself as a 'modern woman', answered:

> There are things that I would expect, being a modern woman, that maybe people wouldn't have expected 50, 60, 70, a 100 years ago. I

> wouldn't live in an environment unless I did have for example running water, unless it was unavoidable. Obviously, I wouldn't choose to live in an environment like that and I see water as a necessity, but also it is very convenient [that I do not have to] worry too much about it, you know wondering where my next glass of water is going to come from or wondering whether how I'm going to wash the clothes or anything like that.... I think, you know, we are talking about [being] 'modern' and 'women' – modern in a convenience sense and as a woman, I think, you know, just being able to have the availability of water in my life is something that I don't really need to think about. So it allows me to think about other things.

In Lauren's narrative, being a woman involves the notion of a special responsibility for water and for the family in general. Being a *modern* woman, however, means that modern water infrastructure frees her from the heavy work of carrying water and the worries of how to access it. According to Lauren's account, it is modern *facilities* that allow her, as a woman, to be emancipated from traditional chores, rather than challenging the understanding of gender roles.

Relating water to pleasure and play was exclusive to the suburbs. Water was spoken of as something that produced well-being and a feeling of being at home (see Kaika, 2005). The soaking of one's body in water is an especially important part of the everyday pleasures that emerge in these narratives. When Carol was asked what she associated with the word water, she answered very quickly: 'Pleasure! Ja, I love water.' Relatedly, Lauren said:

> I would associate it with all good things, with swimming and sparkling oceans and perhaps to go further: serenity I suppose, or you know, I really enjoy swimming so relaxation as well.

Similarly, Karen talked of swimming:

> Swimming is something that I love to do, swimming in the swimming pool, swimming in the sea, swimming in, well anywhere. I love to swim and I love to shower. I shower twice a day.

In narratives from other sites, water is linked to joy (as in Sipho's narrative, discussed below), but in those cases the joy is related to the fact that water is *easily accessible*. In the suburbs, however, the *pleasure and entertainment dimensions* of water are recognized as values of water in their own right, which is an entirely different way of relating to appropriate uses of water.

Such a (water) affluent way of life, and its place-boundedness to the middle-class suburbs, was also a central theme in Mcebo's narrative. He explained that excessive water use was 'part of the whole lifestyle' in those suburbs and that such water use was made possible by the fact that most people there could easily afford water. Mcebo had previously lived in both rural and township areas. He said: 'Whites and us black people who live in the former white suburbs have no respect for water in the African sense ... water is a scarcity in Africa' and he continued, 'we cannot be accusers anymore, we can only be accused by people who are outside the ... suburbs.' Notably, the 'we' in this quotation transforms from referring to a collectivity that was a victim of inequality to that of a community that is to be blamed for it. He explicitly addressed the issue of belonging:

> There is so much in common in a class. Once you are in a particular class [I have] more [in] common with my neighbours than with [Mdlalose] in the township.... There isn't a thing that is as strong as class, class is stronger than the tribal links, stronger than anything. ... I wish to carry this lifestyle and ... and take my house and just boom! [I would place it] inside the township.... Because the things I enjoy there I cannot find them ... the things I enjoy here I want to go with them there but it's impossible to do that. I've lived the farm life and I have lived the township life. So I really miss both ... because of the language, the mood, the food.... You know, the noise, the disorder of some sort, you know the ... *ubuntu*.... It's really nice sometimes when I go there I will go for a hair cut and I end up spending the whole day there and come back in the evening.[6]

Hence, in Mcebo's story, life in suburbia represents individual convenience, but at the same time does not offer a sense of belonging to a community.

Some of these narratives from the well-off suburbs are blurred by feelings of uneasiness and guilt because of how water supports their affluent lifestyles in a context of uneven distribution of the resource. An exploration now follows of how water users from the different sites make sense of such an uneven distribution of water in general and the right to water in particular.

Making sense of the right to water

The Free Basic Water (FBW) policy is a key measure to address water accessibility in South Africa and in eThekwini. As has been mentioned already, the policy is globally renowned and was, in fact, first developed in eThekwini. In their narratives, the water users in the municipality express their views on the right to water and the policy of FBW. While there is a strong discourse about the notion of a universal right to water for basic needs, there is ambivalence in how the narrators speak about what *everyone* has the right to in relation to their understanding of their *own* right. It becomes clear that deprivation, privilege, and social hierarchy play key roles in how they relate to these themes and thereby inform their water subjectivities. First, however, a few words on the relationship between 'good' and 'bad' water (cf. Kaika, 2005)

Water in 'nature', water from the grid, and the right to water

In the water users' narratives, natural sources of water (such as rivers, lakes, and streams) are associated with pollution and sickness. With few exceptions, the general sentiment in the narratives, in relation to such dirty and unhealthy water, is that everyone should have the right to be provided with potable water for household consumption and, additionally, that one might need to change one's own use and behaviour around rivers and other natural sources of water. Hence, rather than seeing water in nature as 'pure' and 'natural', many of the narrators convey that they see such 'natural' water as a possible threat to life and health. Therefore, their stories tell us, water for human consumption requires human intervention in terms of purification processes and the construction of delivery networks, which, furthermore, indicates that practices of fetching water from the river is something that should belong to the past.

Such connotations of water and the need to purify water from natural sources point to another general feature of the stories: the *distinctions* made between humans, animals, and nature. What appears in the narratives is not only the *need* for clean water but also the notion that humans (in contrast to animals) have the *right* to water. For example, it is acknowledged that humans ought not to be left to die of thirst, which is a phenomenon seen as natural in relation to animals. In this sense, humans are spoken of as distinct from both animals and the rest of nature.

In a way, one would think that this acknowledgement of humans as right-bearing individuals could have the political potential to produce common notions of humanity, to connect people, and make visible the similarities between humans. However, the following pages explore how the eThekwini hydromentality, and the way that the right to basic water has been implemented, have opposite effects. According to the voices of the water users, differences between people are actually exacerbated.

FBW and improvement in life conditions

One basic point it is important to make, however, in relation to the FBW policy, is that, for many, it has improved their life conditions. Hence, as a way of biopolitically optimizing the life of populations, it can be deemed a success.

Those who have recently been provided with their FWB in the rural area of Mzinyathi, and who have also been provided with the so-called ground tank, testify to this. The instalment of the tanks means that inhabitants (mostly women and children) no longer have to walk long distances to fetch water and that the water they receive in their yards is clean.

When making sense of the changes that have taken place regarding water access, Sipho, a narrator living in Mzinyathi, took his point of departure as the changes that came with the ending of apartheid. Sipho said:

> We got *a new government that started thinking for the people*, that they must drink healthy water. This water supply came with the new government. Everybody's happy now [that] we got clean, healthy water. (My italics)

And further on in the interview:

> Now we don't carry that water from the borehole or from the river. It changes people's lives. Instead of walking hundreds and hundreds of kilometres, miles, to fetch water, the water is just there, just near.... That water brings joy. It just brings joy.

In Sipho's story water access is associated with the democratic government's *care for the population*, which is central to biopolitical strategies of governing. Water service delivery has a positive connotation, in that government is understood as working to improve people's lives. When Lungi, a young woman also living in Mzinyathi, was asked what she thought of when she heard the word water, she said: 'I think about water those days when we used to fetch water down there by the river. And I hated it. Ja, that is what I think about water.' For Lungi, the newly installed tanks represented a sense of liberation from hard work, a feeling she embraced enthusiastically. Thabisa, another young woman living in Mzinyathi, answered the question of what she thought was the purpose of the tank with: 'The person probably wanted us to have a good life.' A purpose she thought had been fulfilled, she imagined it to be a result of one individual's goodwill rather than part of a whole governing assemblage with a distinct idea of political calculability (cf. Braun and Whatmore, 2010, p. xi).

While many of the narrators in Mzinyathi considered the tanks an improvement, some spoke simultaneously about their worries about interruptions to the water supply that had recently taken place.[7] According to these accounts, the improvement in life conditions had thus created new vulnerabilities, and another type of inconvenience than those experienced in the past. As we shall see further on in this chapter, these interruptions made the inhabitants save water for each other, but it also meant that some (once again) used (polluted) river water for their daily needs. According to the narrators, it was this *unreliability* of whether or not there would be water in the tanks the next day that created the most frustration, rather than the *limitations* on the amount of water. A related problem that emerged in the narratives was that the water users who experienced interruptions did not know what was causing them. When Mrs Zulu was asked why she thought that they did not receive information, she replied, 'I think it's because we are from the rural areas and we are black'. Hence, according to this statement, Mrs Zulu experienced the way that she had been included in the water service delivery system as an element of marginalisation.

In the informal settlement in Crossmoor, the inhabitants spoke appreciatively of the standpipe that had been placed close to the settlement and which guaranteed that the people living in the settlement got access to their FBW. Clairese told me how the community members felt when the standpipe was installed: 'Oh, we were on top of the world! It was a good victory for us to have our own water.'

However, even if the conditions in the settlement had improved considerably, the inconveniences and difficulties of the life people led in the community were common themes in the narratives. The instalment of the tank also involved problematic aspects in terms of the need for communal control of the usage of the water from the standpipe. This will be more closely discussed further on in the chapter, in relation to processes of responsibilization. Relatedly, the narrators' understandings of the tanks in Mzinyathi also have problematic aspects in terms of how, at the same time as improving life conditions and being a symbol of the new citizenship in South Africa, they also represent a reproduction of the unequal distribution of resources under the apartheid era. Such aspects of the implementation of the FBW are discussed below.

FBW and social hierarchies

The narrators spoke about the FBW, and the technologies installed to provide the water, as related to social hierarchies, both within and between communities.

For example, when speaking about the instalment of ground tanks in rural Mzinyathi, Sipho said: 'I think they made [the tank] on the basis of poor people just like us [in] poor communities.'[8] Hence, in his narrative, the tank is simultaneously a representation of the new government's care for the population and a symbol of poverty.

In townships that previously had been connected to the water grid, the introduction of the FBW meant that it was easier to monitor water use and to make the inhabitants pay for their water services (see also, for example, Loftus, 2005). For instance, in areas known for having practised non-payment both during apartheid and after the democratic transition, flow limiters have been installed. These meters can be programmed to cut water access after the FBW amount has been used. They can also allow a flow of larger quantities of water, if the municipality judges that the water user is able to pay for the excess. These particular meters have an effect on the dynamics in the communities in terms of producing particular subjectivities (we will return to this) as well as making distinctions between water users within these communities.

Kamal's narrative testifies to this. Since he was not in possession of a legal contract for his flat, he was not allowed to choose the unlimited option. On the cover of his water meter, the figures he was allowed to access were printed, visible to everyone. Kamal commented: 'It shows everyone that I am struggling.'[9] And indeed, it was common knowledge in the community who had their water restricted. When I arrived and asked community members I met on the street whether they knew somebody who relied on a limited amount of water they could immediately direct me to their flats.

Hence, water governance as a form of biopolitics that steers, administrates, and shapes the lives and lifestyles of different populations, can also be understood as containing a disciplining aspect, in terms of how it simultaneously targets different individuals who, for various reasons, have to be separated out from the rest of the population (for example, illegal tenants and those with large debts). This has the effect of making the individual water user feel stigmatized. In turn, as we will see more closely under the next headline, this singling out of particular individuals has an effect on how other members of the community understand themselves (despite living in an area targeted for these kinds of responsibilizing technologies), namely, as respectable and responsible citizens.

In this particular case, the fact that Kamal and his family had their water restricted also had gendered effects. He told me, seemingly with a bad conscience, that he had blamed his wife for using too much water on those occasions when the family had used up all its water. Moreover, his water situation affected his identity as a father because of how he thought he failed to provide for his family.

In the water users' accounts, the biopolitical claim that everyone should have the right to water for *basic needs* corresponds to governing logics, not only because it embraces the idea of the right to water, but also because it makes a distinction between the water we need for subsistence and uses above that level. Such a distinction relates strongly to the disparities between different populations in the country.

In many narratives, there is ambiguity regarding the understanding of what *everyone* has the right to in relation to their understanding of their *own* right. Karen's narrative is illustrative. She lived in the well-off suburb of Glenwood. When I asked her about what she saw was her right, as a South African citizen, in relation to water, she answered:

> Oh, that's hard in context of the disparities. I don't know how to answer that question. I mean it should be that everyone has the right to drinkable water … potable water inside their house.

Here, Karen's way of bringing up difficulties when defining her right to water as a South African citizen in relation to the disparities in the country, suggests that she found it hard to define a basic level that *all South African citizens* are entitled to. This became especially clear as she problematized the things that she saw as fundamental to an acceptable and liveable standard when it came to *her own* water situation. She said that she could not imagine living her life without the services they were receiving at present:

> If I didn't get water coming into my house the way that we do ... I really would want to go live somewhere else ... I feel very strongly that *I could not personally, given my history* at the age of 42, I could not live if they couldn't, if there wasn't water coming out of the tap. (My italics)[10]

She then seemed to realize that such claims exceeded what she had just asserted that everyone in South Africa should be entitled to. She therefore added:

> I suppose in a way, the way that I'm talking it makes it sound like I think it's my right to have water coming out of my tap if I can pay for it ... and I do: hot and cold water.

During the course of her story, Karen seemed to realize that her understanding of what people have the right to in terms of water was connected to whether they can pay for it or not. While this seemed to make her uneasy, her narrative reflects the view of the dominant discourse, where payment is seen as the distributional means to determine who is eligible for more than their basic water allotment.

Hence, many of these narratives reflect a prevailing view of resource distribution and environmental governance in their constructions of equity, but ambivalently so. While the right to basic water is often straightforwardly acknowledged, the dominant notion of payment as a way to allocate water above the FBW level complicates the water users' narratives about just water distribution. Often, however, in the absence of an alternative, narrators represent payment as a legitimate way of both allocating resources and creating responsible behaviour around water (we will return to this). This is a prevalent theme in the stories of narrators from different communities and with different socio-economic backgrounds.

Resistance against separation between different forms of lives

Even if many water users adopt much of the governing logics of the municipality's hydromentality, some (seemingly both consciously and unconsciously) resist its policies.

As noted in Chapter 4, there are also those in the community who have 'made their own connection' to the municipal water network without permission. According to the municipality, these 'illegal connections', as the municipality calls them, are a widespread problem. Andile, a young man living in Mzinyathi, had such a connection and when speaking of it, he continually referred to the needs and wants of the broader community and to the poor in South Africa:

> I can say *it is a system of showing them that we can access water without their ... without consulting them....* If I am doing this I am showing them that ... there is another way to access water without paying these money ... these laws ... which control us. Which say if you are doing this, you are doing this you are wrong, if you are doing that you are right.... *As we are a community* we don't have many voices.... We are like robots, we are being programmed by these laws which are ... pressing us down. We don't have any view. Because we are not happy with them, they are not favouring *our needs*. Especially the poor people, here in South Africa if you are poor you are going to be poorer, if you are rich you are going to be richer because the laws are favouring the ones who are rich. (My italics)

Regarding the cheap technical solution put in place in the rural areas, he said: 'I'd like everything to be equal, I'd like people to live the same life.'[11] This can be read as resistance against the governing logic of the municipality, in terms of its 'will to divide' (cf. Walters, 2004) through the application of different service levels, policies, and technologies for different individuals and populations.

Illegal connections are common in Ntuzuma and Inanda. Thandiwe's household in Ntuzuma, Lindelani, had connected to the standpipe on the road in this way. She explained: 'We saw at other houses when they were doing it, others have it and [we] connected it inside the house.' In her view, these connections were common practice and she did not understand herself in terms of being an 'illegal' water user. Nor was it a political statement, as it was for Andile. Rather, it was an act that was produced by the everyday need for water together with notions of what are accepted solutions in the community.

From stories such as those related above, it is clear that the idea of communities self-policing 'illegal connections', as put forward by Macleod and Gounden in their interviews (see Chapter 4), is unthinkable in a context where such practices are normalized. The way water users form subjectivities and notions about proper behaviour regarding water is related to common practices in the community rather than to the hydromentality of the EWS. Hence, governing logics cannot be seen as only *illegitimate*; rather, in relation to the water users' everyday lives, they can even be viewed as *irrelevant*. Such stories account for how biopolitical regulation fails because of its insensitivity to context and people's material needs.

The making of (un)sustainable and (ir)responsible water users

Technologies of responsibilization are central to the hydromentality of eThekwini as a means of creating sustainable and responsible water users, as argued in Chapter 4. Therefore, one of the main questions in relation to how water management matters in people's lives from a biopolitical/ governmentality perspective, is how it engages water users in their own governing and invests them with particular self-understandings (Cruikshank, 1999). This can be understood as working through technologies of the self, as water users turn their gaze on themselves (cf. Ettlinger, 2011, p. 539) to become the (ethical) water users they want to be.

This is a question of how water users understand their (moral) obligations in relation to the prescribed code of conduct and how they acknowledge themselves as members of certain populations (Foucault, 1992, p. 27) with certain values and practices attached to water. Central to such research are notions of truth telling/confession (Bowerbank, 1999), guilt (and pleasure), and notions of the (un)just allocation of water resources and (im)proper usage in the construction of themselves as a certain type of water user. This section explores these themes from the perspective of the water users and how they discussed such themes in their narratives. There is a focus on how the water users tell stories of responsibility and sustainability, as these are key themes in the hydromentality of eThekwini that also resonate in the water users' stories.

When looking at questions of responsibility and sustainability in the interview texts, interesting patterns emerge in terms of: how the narrators, and the way they use and think about water, are responsibilized through their interaction with the municipality and water management devices; and how these imaginaries of responsibility and sustainability affect the way water users understand themselves.

Many of the narrators, irrespective of location, told stories of water shortages/droughts/restrictions. There seemed to be a general sentiment in people's stories that water – seen as essential for life/a basic need/a scarce resource – should not be wasted. Frequently, this was exemplified by 'the running tap', either in the bathroom, when brushing one's teeth, or at the standpipe, or from the tap in the kitchen when doing the dishes. Such examples were often used to illustrate the irresponsible behaviour of others. Mrs Shabalala said, for example: 'Others just, they open the tap and put the washing, they open the tap and clean … they leave the water like that.' Such parts of the narratives can be understood as instances of water users producing images of themselves as responsible and sustainable. They also illustrate how established the notion of water as a scarce resource is.

Besides the recognition of water as a scarce resource, the narrators differ in how they narrate the *reasons* why they care/why they *should* care about how much water they use and for what purposes. A discussion of the different rationalities for, and relationships to, the themes of water wastage/saving that appear in the water users' stories follows below. The main logics why the narrators think they should save water are related to the hard work involved in fetching water and/or saving water for others (in the local community) (dominant in the narratives in Mzinyathi and in the informal settlements), to economic reasons (dominant in the township areas) and, lastly, to the need to save water for the sake of 'the environment' (understood in a general sense) (dominant in the suburban areas). It is important to recognize, however, that these are the overriding but not exclusive logics, as there were multiple and coexisting reasons for conserving water in each site.

Narrating frugality and collective (social) responsibility

In rural Mzinyathi where, historically, water access has been limited by the labour required to carry water, 'water wastage' was not a prominent theme in how people talked about water (but there were exceptions). As Gugulethu explained, 'it is [already] in the system', meaning that the work people have had to do to access water has created a sparing way of using it. She continued:

> What can I say: it is in the system, because we grew up not having lots of things. I learned that you should save things if you have that little; you must try and save and not treat it carelessly. And also that water we fetched, we fetched very far from our homes; you would come back with the buckets and you would know that with this bucket I should do lots of things and not waste water.

Elsewhere, such as in Australia, rural experiences of water as a daily struggle have been suggested as one factor behind the willingness to reduce the use of, and to reuse, water (Sofoulis, 2005, p. 457).[12] This is a theme that occurs in many of the narratives from the townships and also (although less commonly) in the suburbs. In turn, this can be seen as a counterargument against the common assumption that the poor are unsustainable in their use of resources, as put forward in early policies of sustainable development (see Chapter 2), as well as featuring as a theme in some of the narratives.

In Mzinyathi, frugality in relation to water use is not only related to the hard work of the inhabitants (mainly women and children), but also to a responsibility for each other, as the newly laid water system was understood to be both limited and unreliable. When asked how she related to 'saving water', Mrs Zulu answered: 'I am trying to save ... so I can give [some water] to the others also.... A person who has water should give to those that don't have any.' Her story features the notion of a limited amount of water and unreliable access along with a feeling that her own use is related to others' access to water in the local community.

According to such a view, water is understood as a common resource that needs to be cared for jointly, particularly after the change from using freely available river water to using piped water from a water utility. Although Mrs Zulu claimed her right to be provided with water (implicitly by an institution such as the municipality or the state), she simultaneously emphasized her own agency and placed responsibility on herself and other community members to secure their own water access. While these effects – especially the unreliability of the system – might be seen as unintended consequences of the hydromentality of eThekwini (more likely to happen there than in other areas because of the system's construction), the stories about how water users handle interruptions to the supply are expressive of a resilient and self-reliant approach to events of scarcity, an approach acknowledged as desirable in policies of sustainable development as well as in water governance. However, it is important to note that resisting the dependence of being connected to (unreliable) water networks through collecting water from other sources, which was suggested as a strategy in some of the narratives, is becoming increasingly risky because of the contamination of water from unprotected sources.

It is also interesting to note how this way of narrating interruptions to water supply was clearly different from the way inhabitants in the suburbs dealt with interruptions. While the narrators in Mzinyathi emphasised their agency in getting water from each other/from other sources of water, the narrators in the suburbs told stories of powerlessness. Karen told how she had experienced a (planned) interruption of water supply; about having to go into other people's houses to bathe and to collect water, which she found both 'irritating and hideous' and as involving 'a different boundary from what [they were] used to'. In such stories, water and water access were closely related to the desire for private space, as already discussed on the basis of water stories in the informal settlement of Crossmoor. Water interruptions were associated with feelings of discomfort and anxiety (Kaika, 2005, p. 68). These stories from the suburbs signalled a feeling of a lack of agency and control over water issues as they not only focused on interruptions but also on the lack of knowledge about, and influence on, the decisions that are being made in terms of water, especially with regard to how water is cleaned and treated. (A concern about) the political future of South Africa seemed to be invested in such stories. In that sense, we can understand these water narratives as a continuous dialogue with both future and past experiences of water, but which, in contrast to experiences in the rural areas, were highly individualized.

Coming back to the issue of collective responsibility, similar logics to those of Mzinyathi also prevailed in the informal settlement of Crossmoor.

As in Mzinyathi, where people also exclusively relied on FBW, water accessibility required 'collective forms of organization and collaboration' (Ahlers and Zwarteveen, 2009) to a greater degree than in communities where people pay for their water.

In the Crossmoor community, the need to steer consumption derives from the understanding that the municipality monitors water use so that it does not exceed the FBW level, calculated on the number of people living in the settlement. This came to the fore when Samuel was asked about the standpipe. He explained:

> One thing I am not happy about is the meter. It would be better if it [the standpipe] was in the land [the settlement] and had a meter to count the amount of water we all use here.... It is hurting [us] because we are told that they [representatives from the municipality] will stop the water if we don't look after the standpipe.... This is counted on us, the meter is there and it is pointing to us.... *It is our responsibility.* What I think is, what they are doing with that meter is that they are monitoring the way water is being used, but what I can say is that *water is not being abused.* (My italics)[13]

Hence, in Samuel's story, the perception of metered water and the threat of cut-offs function as a way of regulating water use in the settlement. Other narrators in the community also explained how a certain limit was calculated for them and that if they exceeded that limit water access was going to be cut off. Hence, as in the case of Mzinyathi, there is an understanding that one's own consumption of water is related to that of others in the community. The notion of the *limitedness* of the resource is present and constitutes an overriding logic in the informants' narratives of their everyday water use.

Because the standpipe was situated on the road downhill from the settlement, the problem of controlling use of the water from the standpipe was amplified. People from other settlements and taxi drivers came there to use the water and the narrators explained that actions were taken against 'outsiders' who used the water. This sometimes gave rise to conflict, Samuel explained, as 'outsiders' did not understand why they could not use the water since the water was provided for free, which sometimes resulted in physical violence. In turn, this required that community members organized themselves around water regulation.

In the narratives from the settlement, the community and its members were spoken of as responsible not only for their own water consumption as a group, but also for others' consumption, which required them to police usage – a task that they do not have the legitimacy to carry out and that jeopardizes their security and hence produces a particular kind of precarity (Butler, 2009).

Narrating economic (ir)responsibility

In the townships of Ntuzuma and Inanda, it was common for the narrators to bring up the hard work that was involved in fetching water from rivers and streams, which they discussed in relation to their present efforts to save water. Mrs Shabalala, for example, talked about the heavy work of carrying water and explained that her present situation was better and that 'we are so happy now we just try just to save'. The overriding logic in the townships when discussing waste waste/saving was economic. Mrs Bongani, who resided in an RDP house in Ntuzuma C together with her son, his wife, and their three children, got her water through the semi pressure system. When asked what she thought about when she saw her meter, she replied:

> It is counting, so it is teaching me to save water.... The mind is always thinking if you are wasting the water. The meter there.... It is saying what you are doing.... Water, you don't just use [it] carelessly.[14]

In Mrs Bongani's narrative, the water meter appeared as an object with a certain form of agency that monitored her behaviour around water so that she used it properly. The overriding rationality for conducting herself in such a manner was economic, as the household had problems covering their expenses. She explained: 'That money we are paying, sometimes you don't even know where you will find [it].'[15] She further emphasized that the household only used water for washing, bathing, cleaning, cooking, washing the dishes, and drinking, and that they did not use the 'metered water' for watering the vegetables in their garden.

In a community consisting of a few blocks of flats in Crossmoor, where flow limiters had recently been installed, the responsibilizing effects of payment, and of the interaction with water technologies and the municipality, emerged very clearly in the narratives. The community was poor and had previously practised non-payment for water services.

Suleilah, who lived there with her mother, her husband, and her child, explained how the community had been invited to choose between restricted and unlimited water supply. Suleilah and her family had chosen the 'unlimited' option and she was positive about the system, as she said that it was easier to keep track of costs when the water bill was separated from rent and electricity.

When Rubi, another resident in the area, was asked why she thought they had installed flow limiters in the community, she replied:

> The reason why they put the flow limiter, you see, people weren't paying for their water. So they had to go for the flow limiters ... Ja, they gave us the flow limiters, because people were owing a lot of money towards water yeah so they started to say ... they gave everyone

> their individual meter, *so whatever they are using they must stand responsible for it*. (My italics)

Suleilah provided reasons for past non-payment for services, partly that households could not afford it, but there were other logics, such as it was a resistance strategy against the apartheid state (see also, for example, Desai, 2002; Freund, 2010). In the particular area where Suleilah and Rubi lived, the residents wanted to prove a point in order to get ownership of their homes (which the inhabitants were offered subsequently).

Both Suleilah and Rubi constructed themselves as responsible through their willingness and ability to pay for water. Payment had hence become a way for them to prove themselves worthy of inclusion in South African society, thus providing them with a particular sense of self as worthy citizens. Suleilah said:

> I always told them ... I wanted the unlimited ... and I didn't have a problem at all because for me it's normal I take care of it ... I know where my money is going.[16]

In a similar vein Rubi told me that she felt "proud" of paying her water bill. Suleilah said:

> We are very, very conscious now because we have this, this new system that we actually have to pay for water. So we have unlimited water, but we take care of our water. There is no more ... free for all ... but that is ok. We limit ourselves, we restrict ourselves to a certain thing.[17]

Being able to limit and restrict themselves in relation to water becomes a virtue for a whole community. Such virtues are important parts of environmental discourse in terms of how individual subjects work on themselves in order to become 'green' (see Bowerbank, 1999). As discussed in Chapter 3, such acts of narrativization can in themselves be seen as part of a self-transformation to become more environmentally minded (Mikulak, 2013). From this point of view, Suleilah actively shaped her view of herself as sustainable in the moment of telling her story.

Furthermore, in Rubi's story, there was a certain sentiment that if people do not conduct themselves properly, they can be seen as having *brought about their own misery*, which has been a common theme in how the poor and their resource use have been moralized throughout environment and resource use discourses, as discussed in Chapter 3. Rubi developed this theme when queried about how she viewed the tension between the *responsibility* to pay and the *ability* to pay. She responded:

> You see those who got a big part of arrears right ... I think that they should go to the department.... Make an agreement that how much

> they can afford to pay towards that amount that they are owing. Because in case it comes to a stage where they are going to come and cut off the water, what are you going to do then? They rather make the agreements now than when the damage is done. That's how I feel. They rather make proper agreements now.

According to Rubi's logic, people who are struggling with payment and who are burdened with arrears that they owe to the municipality should apply for debt relief and enter a contract with the municipality. Hence, Rubi actively advocated a certain type of behaviour that was in line with the logics of the municipality.

Furthermore, Rubi had concerns about what happens if people do not pay for their water. It was not a concern about people being cut off from supply, but rather a concern about the municipality not taking proper action *against them*. She said that she did not understand why people refused to pay now that flow limiters had been installed. She elaborated:

> They wanted the unlimited, fair enough ... they got the unlimited, yet they were in arrears and still they never kept their promises.... They are still not paying, then why should they have the unlimited? ... If you give them like the [repeats words] 300 litres, it's for free right. And whatever that is after that you pay for it. So once that is finished, why don't they disconnect it? Because they are still not even paying for the water. That's what's more like, you know, upsetting.

In Rubi's way of speaking about the flow limiters, the act of choosing between limited and unlimited water becomes responsibilizing in the sense that it represents an approval of governing rationales. It also produces a conflict line between those who pay and those who do not pay, as she understands that those who pay are being 'punished' because of those who do not pay. Rubi also claimed that the municipality was not doing its work properly since its staff, as she understood it, did not disconnect those who did not pay. In her presentation of the problems of water service delivery in the community, there is thus a need for a government that *disciplines* the water users in a more detailed manner.

In the narratives discussed here, the flow limiters can be understood as part of an assemblage that makes individual water users 'responsible through their individual choices for themselves and those to whom they owe allegiance' (Rose, 1996, p. 57). The water users are made active in their own government; their choices – to either restrict water or not – place responsibility on themselves if they are unable to pay their bills or if water is cut off (see also von Schnitzler, 2008). As is clear in the narratives, such a way of governing is successful in investing the water users with a positive self-understanding.

Narrating ecological (ir)responsibility

The stories around the new meters also contain ideas about *sustainability*. In Suleilah's story there is a notion of water as a scarce resource that is 'not to be wasted at all'. In turn, such desirable conduct is related to payment. This is not only a dominant view in global environmental discourse and in the hydromentality in eThekwini, but also a recurrent theme in the water users' stories.

Zinhle, living in Inanda, narrated payment as the way of making people care about how much water they are using. She constructed herself as 'responsible' and 'sustainable' in contrast to others. She said that she was afraid that water would run out because 'people just waste water' and that there was a need to teach people not to do so. She fully adopted payment as a measure to responsibilize people in their water use. She said: ''Cos *it makes people think before they use....* They mustn't waste, I think it's better to pay for water.'

In Zinhle's way of speaking, the proper and frugal usage of water thus appears as something that people must be taught, as her general understanding is that people are wasteful. Furthermore, Zinhle's narrative is illustrative of how many of the water users construct themselves as responsible but attribute irresponsible and unsustainable water usage to others. In this way, the narratives function so as to construct an ethical (responsible and sustainable) self, which, in turn, requires an image of the *un*ethical (irresponsible and unsustainable) water user.

However, the strength of the discourse of payment can be illustrated by the fact that it also produces narratives of the self as *unsustainable* in situations where water users do not pay for water. This is mirrored in the way that Thandiwe, who accessed it from an 'illegal connection', talked about her own use of water. Thandiwe was clearly governed by a strong discourse of payment in the way she perceived people like herself who did not pay for water. She constructed herself as both irresponsible and unsustainable: 'People that get [water] for free ... they waste it, us people that get water for free most of the time we don't use it appropriately.'

In the suburbs, the theme of water waste was prevalent. To a large extent responsibility was spoken of in terms of 'the environment' rather than in terms of economy. In such narratives, ideas of water scarcity governed notions of responsibility and sustainability. Karen, living in Glenwood, spoke of her water meter in terms of water waste:

> It just goes [tjick-tjick-tjick-tjick-tjick] and ... I feel a bit of a panic and I suppose it is related to some kind of environmental sense of the water crisis that our world is in and ... worrying about that, and worrying what's going to happen to our world because our water supply, I think is finite, and we don't treat it like that. And so I do have a little bit of a panic about: ah! You are wasting water![18]

Hence, the meter – through the sounds it produces – evokes feelings of anxiety in Karen. This worry is a recurrent theme in her talk about water, and Karen spoke of what she thought the ignorance of the water issue would involve: 'I think it might be in some ways kind of the end of our world as we know it, the water crises.' In Karen's narrative, water issues are strongly associated with future threats to her way of life and she thinks that both she and society in general need to become more aware of sustainable ways of life. Karen's feelings towards her water meter highlight the significance of the technological devices in mediating understandings of our human–nature relationships; the ticking of the meter is a measurement of the household's depletion of the earth's water resources.

While it was common in the townships and the informal settlement in Crossmoor for the narrators to bring up the example of children playing with water to illustrate irresponsible behaviour and water wastage (as shown above), there was, in the suburbs, a more positive attitude to playing with water. In these areas, water had an entertainment value, which was exclusive to this particular site.

When I specifically asked Lauren whether she had reflected on the water needed for filling up the swimming pool, she answered that she did not consider that as a 'waste' but rather a 'very good use of water' because of the 'entertainment value' and because it was 'the best thing in the world just to submerge yourself in the water'.

Thus, in suburban areas water is used for additional purposes, such as for filling swimming pools and for leisure gardens, compared to other parts of the municipality. Karen nevertheless framed gardening in terms of environmental consciousness. She said: 'The last few years we became very environmentally conscious ... for example we have planted all indigenous plants in our garden so that we don't have to water as much.'[19]

This means that while the narrators in the suburbs are those who are most clearly constructing themselves as environmentally conscious, they are simultaneously the ones who are using the most water. This divergence, between environmental subjectivity and practice, highlights the uncertain material effects of governing technologies, as they work through the governing of the self; the (environmental) subjectivities that are produced do not necessarily relate to how (much) water is used. A possible interpretation of this is that infrastructure (together with other factors) places limitations on the agency of the narrators with regard to their becoming a particular kind of water user. In the suburbs, for example, the households' water infrastructure constrains the frugality that can be achieved with regard to water use, as water closets and big bathtubs require relatively large quantities of water. Therefore, as Sofoulis has highlighted, we should not assume that:

> practice is a direct product of values, or that information pumped into a person changes their behaviour, this perspective suggests values

> could change as a result of practice, and reminds us that practices and values can operate somewhat independently.
>
> (2005, pp. 450–451)

Hence, the narratives should not be seen as instances where water users *mediate* their actual conduct in relation to water (but rather how they produce themselves in relation to the conduct of conduct, that is, the hydromentality). As theories of materiality remind us (Braun 2005; Braun and Whatmore, 2010; Head and Gibson, 2012; Shaw, 2012; Whatmore, 2006), when aiming to understand the factors that are relevant to the production of behaviour around water use, we need to take into account how a whole assemblage – the water network, meters, policies, billing systems, domestic water infrastructure, and gardens – enables and constrains the agency of the water user. Such a focus is important to understanding the relationship between technologies of the self – the work people do on themselves to become the (ethical) water users they want to be – and the surrounding factors that are part of producing their everyday routines around water.

In Carol's narrative, a sense of a discrepancy between awareness and practice emerges. In the beginning of the interview, she said:

> Look, I'm vaguely water conscious. I always ..., growing up as a child my parents were very conscious of excess. I am aware. I married into excess ... I'm married into affluence so I still remain very conscious of excess and wastage and all those sorts of things so ja, I'm aware of it.

Carol lived in a big house in Berea (for instance, it contained eight bathrooms). When asked about how much the household approximately spent on water each month, she replied (almost whispering): 'I haven't a clue.' She asked one member of the house staff to find a bill for her. The conversation went as follows:

STAFF: *[talks as she comes closer]* on average ehm one-four-one-five
CAROL: Our water bill? *[sounds surprised]*
STAFF: [name of a male person] says it's because of the irrigations is the same as *[one word inaudible; she speaks from a distance]*
CAROL: Gee!
SOFIE: How much is it?
CAROL: One thousand four hundred rands a month
SOFIE: Only for water?
CAROL: That's just water?
STAFF: Yeah.
CAROL: Not water *and* electricity [*her own emphasis*]?
STAFF: No *[inaudible; she is speaking from a distance] [clears her throat]* electricity is about another 1,000 rand.

CAROL: Ja, we irrigate our garden, we have a swimming pool, we irrigate our garden. How many kilolitres of water is it?

STAFF: Between 90 and a 100 kilolitres. *[Clears her throat]*

CAROL: Gee.

SOFIE: Mm.

CAROL: Ja.

SOFIE: Ja.

CAROL: So there you are ... excess.

...

CAROL: You know you are taking it for granted, but it doesn't stop you. You still got irrigation systems going and swimming pools and deep baths and....

Our talk about water clearly affected her in an emotional way, as she said right before I left the house: 'And once you leave oh Jesus I have to think about water. Why are you here!? *[Laughter]* Panic!' Such feelings of guilt in relation to water consumption have punctuated the narratives from the suburbs. As this is a recurring theme at this site, it signals that these water users present their own (excessive) consumption as a *problem*. By producing an image of being guilty and confessing that they are not being careful enough about how much water they are using, they present themselves as ethical water users, who, despite how they act, ultimately know right from wrong. Such a discrepancy, between the notion of the virtue of conserving water and the water users' practices, is a beneficial outcome for the municipality. As Sofoulis has noted, the more water users are saving water (in this case, those who are paying), the less money the utility makes (Sofoulis, 2005, p. 455).

Stories from Raising the Citizens' Voice

On the other side of the coin, the less water used by those who rely on FBW, the less expensive it becomes for the utility to provide this water. Among the 64 narrators interviewed, a handful (only from the poor communities) attended the water training project Raising the Citizens' Voice. The project made a significant mark on the participants' narratives, especially in terms of how they narrated their knowledge, rights, and responsibilities in relation to water issues.

Knowledge about the processes involved in getting the water to the users had created awareness of the need to *save water*. Samuel said:

> Water – life. Life, very important, very important that everyone must have access to water. That is why I say I am happy to know more about how water comes to us. Because most of the time we have been using water not knowing what is happening. Actually when water is cleaned and sent to us now that we know. We are happy and we want to save the water. That's another reason why I say I must keep

> managing within the free water and add storm water to do other things that need water.

What Samuel spoke about here was a hypothetical situation in which he lives in a house of his own with an individual water access point (in the settlement, no one pays for water, as outlined previously). Here, learning about the billing system and about how to save water seemed important in the construction of a hopeful future.

Just like Samuel, Ayanda, who lived in Mzinyathi, connected water training to water saving. When asked what she thought of when she heard the word water, she said: 'I just think of eThekwini Municipality. They are training people to learn how to save water and I just think of, in my house, how to save water in my house.' According to Ayanda, there was a need for such initiatives, as she claimed that 'there is no information how to save water [elsewhere]'. Also for Ayanda, who at the time of the interview accessed water through the ground tank, water saving was projected as an image of a bright future in which she would have a big house and a maid whom she would instruct to save water. Ayanda said:

> I would tell her: 'My sister, don't waste water' *[Laughing]*... And I would tell my child 'don't play with water, don't [flush] the toilet every time'. Because they are wasting water, the children, they are flushing all the time, flush, flush, flush.

Individual responsibility and the importance of paying for water

Through his narrative, Samuel produced an image of the proper way of living individually and taking *individual* responsibility for water, while there was (as has been discussed above) little space for individuality in the settlement. When I inquired what Samuel thought of the connection between water and the good life, he reflected specifically on this. If houses were to be built for the people in the settlement, Samuel said:

> [then] we *can choose as individuals the ways of saving water*, and how to use water and how to do everything our way. Not like here, we are all using water.... Some people will cry of the way that water is used, some people use more water. Some will go down there and just pull the tap down and leave the water running [They will do their] washing while the water is running. When they arrange clothes after washing, they just pull this thing down and stand and talk, while the water is going. It's causing problems. People are reckless; reckless like now *we are not paying.* (My italics)

Samuel's account is similar to two of the other participants' stories of the water-training project. Both narrators focused on how they, due to the training, understood the importance of paying for water. Cebisa, living in the informal settlement in Crossmoor, said:

> I know water is important, but I didn't know how to ... how can I use water. I've learned a lot that I didn't know [that] water is very expensive. I didn't know, now I know. Because I am not paying for water. I can't use water carelessly, because somebody is paying for the water.

Similarly, Aneesa, living in Westcliff, Chatsworth, explained:

> When we go for the workshop and all.... Then I realized that how much it takes for us to get that water in our taps. We say it's free, OK, God gave us the water and should be free. But then to purify that water and to bring it into our taps, that takes money.

As in Samuel's narrative above, the training seems to have provided Aneesa with objective facts about water, giving her insights which put previous understandings of water in a different perspective. Here, the argument that water is God-given loses its merits in relation to the idea of (non)payment. In turn, this means that payment could be seen as a legitimate mechanism in water service delivery regardless of people's understanding of what water is and from whom, where, or what it comes, or to whom it belongs.

It seems that for the participants quoted above, the project has produced particular ways of knowing about and relating to water and water use. In this sense, the water-training project can be understood as an efficient technology in the governing of water users (as claimed by the EWS managers). However, as has been shown in this chapter, the way people speak about water, and their perceptions and values, do not necessarily result in changing practices and therefore we cannot know the material and practical effects of such government.

Conclusion

This chapter has provided perspectives on the biopolitical effects of the hydromentality of eThekwini. The water users' 'bionarratives' have shown how water is constitutive of the lifestyles they lead (or have led). Water and water technologies have spurred stories about lifestyles characterized, for instance, by: fetching water from the river; about 'town' lifestyles that are convenient but require payment for water services (which for many involve a notion of life as a struggle); about (water) affluent lifestyles where the pleasure of (and play with) water is a use in its own right; and about lifestyles in informal settlements where water conditions can be understood as productive of a lifestyle that is degrading to human life. The chapter also provided examples of how water and water technologies are productive of subjectivities and are related to social hierarchies. This is especially clear in narratives on the right to water and the FBW. Here, we have learnt not only that policy makes a distinction between different forms of lives, but also that this resonates in the water users' stories.

The exploration of the water narratives shows how the ways in which different technological devices and service levels have been applied to suit different areas, populations, and individuals, play a role in making distinctions both *between* and *within* communities. Water users in different areas bear different responsibilities in relation to water, they carry out different types of work, and take different precautions to safeguard access to it. This seems to consolidate the disconnectedness of the different lives in the communities of eThewkini. We have also seen how humans are separated from the rest of nature and how water in nature is understood as polluted and dangerous (bad water) in contrast to the purified water from the grid (good water).

Because of water's immense importance both for survival and for the way that people live their lives, good quality of water and good access to water are linked to images of the good life and a good society. Water is, in this regard, central to the legitimacy of the South African state; the narrators' relationship to the (ANC) government is reflected in their accounts of the changes in water access in relation to the past, as well as in their experiences of the present and their expectations for the future. Projections of the (water) future are especially instructive of the narrators' understanding of the political direction of the country. Based on their understanding of achievements thus far, the image of the South African (water) future that is drawn up in the narratives is a future in which access to money dominates the possibility of being able to live a good life (in terms of water).

Furthermore, the chapter has shown how the production of (ir)responsible and (un)sustainable lives and lifestyles is constructed as highly dependent on issues of class, race, and gender. In the last part of the chapter, the water-training project illustrated how dominant ways of understanding water and water issues mark the water users' stories, especially in terms of the need to take individual responsibility for water and to pay for water services. Such constructions of responsibility and sustainability can be noted particularly in the stories around the newly installed flow limiters in Crossmoor, and the accounts given by those water users who attended the water training. In their stories, those who do not act accordingly are written off as creating their own misery. In this sense, governing logics have permeated the water users' understanding of what is considered appropriate behaviour. Importantly, as has been shown above, these processes of governing are uncertain and can take the shape of both resistance against and compliance with governing rationales.

Notes

1 This is the narrator's real name, as he preferred to be referred to in this way.
2 These removals and the lack of compensation are also documented elsewhere; see, for example, *Flooded and Forgotten: The Story of Inanda Dam Displacement* (2009) and Scott and Diab (1989).
3 On 18 January 2012 I received an email from O'Brien in which he wrote that the South African government had, in October 2011, informed the community that they had decided to pay them compensation of ZAR56 million and also build houses for them.
4 Part of this quotation appeared in Hellberg (2014).
5 During the fieldwork ZAR1 was equivalent to around US$0.1.
6 Parts of this quotation and excerpts from Mcebo's narrative have appeared in Hellberg (2014, 2017).
7 The EWS managers explained that these occurred because of the hilly landscape in combination with low water pressure.
8 This quotation appeared in Hellberg (2014, 2017).
9 This quotation appeared in Hellberg (2014).
10 This quotation appeared in Hellberg (2014).
11 This quotation appeared in Hellberg (2017).
12 In Sofoulis' study (2005), there was a positive correlation between 'rural experiences' and the willingness to take part in DIY recycling efforts.
13 Part of this quotation appeared in Hellberg (2014).
14 This quotation appeared in Hellberg (2014).
15 This quotation appeared in Hellberg (2014).
16 This quotation appeared in Hellberg (2014).
17 This quotation appeared in Hellberg (2014).
18 This quotation appeared in Hellberg (2014).
19 This quotation appeared in Hellberg (2014).

References

Agamben, G. (1998) *Homo Sacer: Sovereign Power and Bare Life*. Stanford, CA: Stanford University Press.

Ahlers, R. and Zwarteveen, M. (2009) 'The water question in feminism: water control and gender inequalities in a neoliberal-era', *Gender, Place and Culture: A Journal of Feminist Geography*, 16(4), 409–426.

Bakker, K. (2012) 'Water: Political, Biopolitical, Material', *Social Studies of Science*, 42(4), 616–623.

Bowerbank, S. (1999) 'Nature Writing as Self-Technology', in Darier, E. (ed.) *Discourses of the Environment*. Malden, MA: Blackwell Publishers.

Braun, B. (2005) 'Environmental issues: writing a more-than-human urban geography', *Progress in Human Geography*, 29(5), 635–650.

Braun, B. and Whatmore, S. (eds.) (2010) *Political Matter: Technoscience, Democracy, and Public Life*. Minneapolis, MN: University of Minnesota Press.

Butler, J. (1997) *The Psychic Life of Power: Theories in Subjection*. Stanford, CA: Stanford University Press.

Butler, J. (2009) *Frames of War: When Is Life Grievable?* London: Verso.

Carr, D. (1997) 'Narrative and the Real World: An Argument for Continuity', in Hinchman, L. P. and Hinchman, S. K. (eds.) *Memory, Identity, Community: The Idea of Narrative in Human Sciences*. Albany, NY: State University of New York Press.

Cruikshank, B. (1999) *The Will to Empower: Democratic Citizens and Other subjects*. Ithaca, NY: Cornell University Press.

Desai, A. (2002) *We are the Poors: Community Struggles in Post-Apartheid South Africa*. New York, NY: Monthly Review Press.

Elliott, J. (2005) *Using Narrative in Social Research: Qualitative and Quantitative Approaches*. Los Angeles, CA: Sage.

Ettlinger, N. (2011) 'Governmentality as Epistemology', *Annals of the Association of American Geographers*, 101(3), 537–560.

Flooded and Forgotten: The Story of Inanda Dam Displacement (2009) www.youtube.com/watch?v=3vox53e10Lg, [accessed 22 December 2014].

Foucault, M. (1992 [1984]) *The Use of Pleasure: History of Sexuality II*, Harmondsworth: Penguin.

Foucault, M. (2000) *Essential Works of Foucault 1954–1984*. Rabinow, P. (series ed.). Vol. 3 [Power], Faubion, J. D. (ed.). New York, NY: The New Press.

Freund, B. (2010) 'Is there such thing as a Post-apartheid City?', *Urban Forum*, 21, 283–298.

Gibbs, L. M. (2006) 'Valuing Water: Variability and the Lake Eyre Basin, Central Australia', *Australian Geographer*, 37(1), 73–85.

Gibbs, L. M. (2009) 'Water places: cultural, social and more-than-human geographies of nature', *Scottish Geographical Journal*, 125(3–4), 361–369.

Hall, S. (2008 [1996]) 'Who Needs "Identity"?', in Hall, S. and du Gay, P. (eds.) *Questions of Cultural Identity*. Los Angeles, CA: Sage.

Hansson, S., Hellberg, S. and Stern, M. (eds.) (2015) *Studying the Agency of Being Governed: Methodological reflections*. London, UK: Routledge.

Head, L. and Gibson, C. (2012) 'Becoming differently modern: Geographic contributions to a generative climate politics', *Progress in Human Geography*, 36(6), 699–714.

Hellberg, S. (2014) 'Water, life and politics: Exploring the contested case of eThekwini municipality through a governmentality lens', *Geoforum*, 56, 226–326.

Hellberg, S. (2015) 'Studying the governing of lives through bio-narratives', in Hansson, S., Hellberg, S., and Stern, M. (eds.) *Studying the Agency of Being Governed: Methodological reflections*. London, UK: Routledge.

Hellberg, S. (2017) 'Water for survival, water for pleasure: A biopolitical perspective on the social sustainability of the basic water agenda', *Water Alternatives*, 10(1), 65–80.

Hinchman, L. P. and Hinchman, D. K. (eds.) (1997) *Memory, Identity, Community: The Idea of Narrative in Human Sciences*. New York, NY: State University of New York Press.

Johansson, A. (2005) *Narrativ teori och metod: med livsberättelsen i fokus*. Lund, SE: Studentlitteratur.

Kaika, M. (2005) *City of Flows: Modernity, Nature and the City*. London, UK: Routledge.

Linton, J. (2010) *What is Water? The History of a Modern Abstraction*. Vancouver, BC: UBS Press.

Loftus, A. (2005) ' "Free Water" as Commodity: The Paradoxes of Durban's Water Service Transformations', in McDonald, D. A. and Ruiters, G. (eds.) *The Age of Commodity: Water Privatization in Southern Africa*. London, UK: Earthscan.

Mikulak, M. (2013) *The Politics of the Pantry: Stories, food and social change*. Montreal: McGill-Queen's University Press.

Plonowska Ziarek, E. (2012) 'Bare Life', in Sussman, H. (ed.) *Impasses of the Post-Global: Theory in the era of climate change,* Vol. 2. Ann Arbor: Open Humanities Press.

Ricoeur, P. (1991) 'The Human Experience of Time and Narrative', in Valdés, M. J. (ed.) *The Ricoeur Reader: Reflection and Imagination.* New York, NY: Harvester Wheatsheaf.

Rose, N. (1996) 'Governing "advanced" liberal democracies', in Barry, A., Osborne, T., and Rose, N. (eds.) *Foucault and Political Reason. Liberalism, neo-liberalism and rationalities of government.* Chicago, IL: The University of Chicago Press.

Rutherford, S. (2007) 'Green governmentality: insights and opportunities in the study of nature's rule', *Progress in Human Geography*, 31(3), 291–307.

Shaw, I. G. R. (2012) 'Towards an evental geography', *Progress in Human Geography*, 36(5), 613–627.

Scott, D. and Diab, R. D. (1989) 'Inanda Dam: a case study of the social impacts of infrastructural development in the South African context', *International Journal of Environmental Studies*, 34(1–2), 43–55.

Sofoulis, Z. (2005) 'Big Water, Everyday Water: A Sociotechnical Perspective', *Continuum: Journal of Media and Cultural Studies*, 19(4), 445–463.

Stern, M. (2005) *Naming Security – Constructing Identity: 'Mayan-women' in Guatemala on the eve of 'peace'.* Manchester, UK: Manchester University Press.

Stern, M. (2006) 'Racism, sexism, classism, and much more: reading security-identity in marginalized sites', in Ackerly, B. A., Stern, M., and True, J. (eds.) *Feminist Methodologies for International Relations.* Cambridge, UK: Cambridge University Press.

Strang, V. (2004) *The Meaning of Water.* Oxford, UK: Berg Publishers.

Sultana, F. (2009) 'Fluid lives: subjectivities, gender and water in rural Bangladesh', *Gender, Place and Culture*, 16(4), 427–444.

von Schnitzler, A. (2008) 'Citizenship Prepaid: Water, Calculability, and Techno-Politics in South Africa', *Journal of Southern Africa Studies*, 34(4).

Walters, W. (2004) 'Secure Borders, Safe Haven, Domopolitics', *Citizenship Studies*, 8 (3): 237–260.

Whatmore, S. (2006) 'Materialist Returns: Practising cultural geography in and for a more-than-human world', *Cultural Geographies*, 13(4), 600–609.

White, H. (1987) *The Content of the Form: Narrative Discourse and Historical Representation.* Baltimore, MD: The Johns Hopkins University Press.

Wrangel, C. (2014) 'Hope in a time of catastrophe? Resilience and future of bare life', *Resilience: International Policies, Practices and Discourses*, 2(3), 183–194.

6 The biopolitics of water

Introduction

This book has argued that it is fruitful to analyse water governance from a biopolitical perspective, and that we can understand GWG and the particular hydromentality of eThekwini as a form of water governance that is productive of biopolitical distinctions both between human and non-human lives. The book has also explored the biopolitical logics of global and local water governance, and their effects in terms of how they contribute to constructing lifestyles and notions of life, sustainability, and responsibility.

Thus, the framework used in this book has allowed an analysis of how water and water technologies not only shape people's water use and access, but also play a part in making people's lives (un)bearable and (un)liveable (cf. Butler, 2004) and produce particular understandings of ethical water users, citizens, and humans. It has produced insights into how water and society make each other (Linton, 2010). It has also provided perspectives on how we can understand water not only as biopolitical (Bakker, 2010, 2012) but also how the biopolitics of water produces effects in the intersection between the multiple uses and meanings of water and its governance in relation to people's lives.

This final chapter outlines some of the main theoretical, empirical, and methodological conclusions that can be drawn from what has been presented in this book and how this contributes to the vibrant field of hydropolitics. First, it gives an overview of what a biopolitical perspective on environmental and resource governance entails. Then it summarizes the insights gained in relation to the notion of hydromentalities, with regard not only to the main water paradigms but also to specific cases around the world. Next, it explores the main findings from the study on the biopolitics of the specific hydromentality of eThekwini, while also fleshing out what a methodology, such as the one employed in this book, means for the study of water governance in general and hydromentalities in particular. This is then followed by some concluding remarks on neoliberal biopolitics and how we can make sense of a 'beyond' of such governance.

Environmental and water governance as biopolitical

This book has applied a framework informed by biopolitics literature within the field of green governmentality to water governance. A specific engagement with the Foucauldian notion of biopolitics draws attention to how populations are constructed and governed, and also, by drawing on biopolitical theorists such as Georgio Agamben (1998) and Cary Wolfe (2012), how distinctions are made between different forms of lives, both human and non-human.

This book has placed an emphasis on the way that populations have been problematized in relation to their use of resources and their care for the environment. It has made explicit the genealogy of liberal political rationalities surrounding the, so-called, population–resource nexus. This trajectory of the nexus has exposed how 'the Malthus effect' (Dean, 2015) – which, interestingly, has been neglected in Foucault's scholarship – has (re)produced the idea of overpopulation and the centrality of scarcity in how different populations are thought about and distinguished between. In this context, two things are of the utmost importance to the argument in this book.

First, Malthusian logics modified the links between the biopolitics of the population and liberal economic rationality. According to Malthus, there are *certain* populations, that is, the poor and the 'underdeveloped', that should be viewed as problematic because of their lack of futurity in their relation to resource use. This, in turn, builds on a colonial hierarchy between 'savage' and 'civilized' life that was central to Malthus' population theses. In this way, Malthus introduced a caesura into the biopolitical conceptualization of the population (Tellmann, 2013, p. 142) that entails distinguishing between different forms of lives. It has been argued throughout this book that such constructions of different populations and their relation to resource use are still part of environmental and resource discourses today, although they use a different terminology.

Second, scarcity, in this context, is not only a concept that describes a situation in which demand is greater than the availability of resources, but it also functions as a 'regulatory device' (Tellmann, 2013), or in other words, a governing technique, aimed at instilling a sense of limitation, and a move towards acting 'as if' scarcity continually prevails for the chance of a sustainable future (ibid.).

Such an understanding of the role of scarcity, and the persistence of Malthus' logics throughout two centuries (Dean, 2015), relate to, and explain, why moralism is a central component of discourses around sustainable development and care for the environment. In turn, such moralism is related to how subjects work on themselves through techniques of the self, which is utilized in neoliberal strategies of governing, as was empirically explored in the case of eThekwini.

In the light of the above – that is, how liberal population discourse distinguishes between different forms of lives and, in the process, how certain populations are constructed as a risk to a sustainable future – basic needs, which are at the core of sustainable development, become pivotal for understanding the biopolitics of such a governance discourse.

A biopolitical analysis of the basic needs approach in general, and of the right to basic water in particular, shows that this approach functions in three, interrelated, ways. Basic needs are a way of guaranteeing the life and health of the most vulnerable populations; they are a way of optimizing life at a level that does not lead to 'overconsumption', that is, in a water context, a consumption of water above the threshold of water for essential needs without paying for it. Basic needs can also be viewed as an intervention made so that market rationalities will function. In the case of water, this means that while granting the poor their basic water, commodification of the resource can continue above that level. The effect of such a way of distributing water resources is that it produces a perpetual distinction between those who rely on water for their survival and those who can access more water if they can pay for it. Taking into account the importance of water in all aspects of life, this distinction is simultaneously productive of a difference between a mere life and a good life. Thus, what this analysis suggests is that what is at stake in a biopolitical analysis of present-day water governance, in a sustainable development regime, is the production of different forms of lives and lifestyles. Such governing techniques stand in direct opposition to a politics aimed at promoting social sustainability. As I have argued elsewhere (Hellberg, 2017), improved access to water for those who have previously been without it is important, but it is not enough to produce social sustainability. Social sustainability also demands that we consider the effects of water management on the *relationship* between different populations and individuals in society.

Moreover, a prerequisite for contemporary water governance is the distinction that is made between humans and the rest of nature. Despite the efforts to reconcile different values of water in global water governance, the meaning of ecological sustainability in the policy documents analysed in this book is restricted to a means to meet basic human needs and to aspire to better lifestyles. Hence, 'nature' and forms of life other than human have emerged as a resource base for economic development and human well-being. In this process, only human life is intrinsically valued and the idea of 'nature', or an outside of human existence, almost entirely disappears.

Thus, the attempts of contemporary water governance to govern not only human life, but also other forms of life, make it characteristically biopolitical in the ways that it governs and distinguishes between humans and the rest of nature, and between different populations and their access to a vital resource that affects survival as well as lifestyles and the possibilities of living a good life. Notably, water and biopolitics are concepts that converge. They both centre on 'life' and can both be viewed as tools for a 'transformation of life' (cf. Foucault in Luke, 1999, p. 142). In that sense, regulation of access to water is a perfect biopolitical mechanism. Hence, global water governance is part of an assemblage that has as a purpose (as well as being an essential tool) to regulate and administer 'all the conditions of life' (Rutherford, 1999, p. 45).

Accordingly, the perspective on biopolitics that resonates with the dynamics in sustainable development discourse, and in water governance policy and practice, is one that makes the population and its relation to the resources on which it depends for its survival and health, a central problem. Such an understanding of biopolitics is in accordance with a Foucauldian understanding.

However, because Foucault failed to engage with the caesura that Malthus introduced into the way that populations are understood, he also failed to problematize the colonial hierarchizations of how different populations are conceptualized and targeted in biopolitical governing strategies. Such distinctions still mark the way that sustainable development and water resource governance construct different populations in relation to futurity and the possibility that they will become sustainable. For the purposes of understanding the dynamics in environmental and resource governance, addressing this lacuna in Foucault's biopolitical theories is of great importance. In this context, Agamben's theories aid the process of problematizing how distinctions are made between different forms of lives.

Thus, rather than a fear of the other, against whom we protect ourselves through immunization (Esposito, 2008), the biopolitical dispositif in sustainable development discourse and in water governance, works so as to produce, and make appear natural, the hierarchization of different forms of lives (both human and non-human). Hierarchizations that continue to be based on colonial understandings of what a 'proper' life is (cf. Agamben, 1998). In this way, the life chance divide between rich and poor populations is sustained (Duffield, 2007; Hellberg and Knutsson, 2016) and the privileges and lifestyles of the middle classes of the Global North are kept intact.

Hydromentalities

This book has not only engaged with the concept of biopolitics but also with governmentality.

Based on the latter, the book has proposed the concept of hydromentality to capture the rationalities and ways of thinking that underpin water governance, and the practical techniques through which water use and water users are governed. Such an understanding makes the concept useful for analysing water governance in particular national, regional, or local contexts.

This book has attempted to describe two main water paradigms in hydromentality terms.

The first is the *hydraulic mission*, a paradigm that took many different forms and expressions. Generally, in this paradigm water was understood as a resource for building the nation state, its agriculture and industry, and for creating modern citizens. Its ultimate symbols in the public and the domestic spheres were the dam and the private bathroom, respectively (Bakker, 2012). In this paradigm: water governance was centralized and its top-down decision-making was heavily dependent on the technical expertise of engineers; and roles and identities were established in relation to water along gender, class, and race lines. These distinctions, in turn, were connected: to the division of labour in relation to water (men constructed dams, and hence produced the 'good' piped and clean water, and women used water in the household [Kaika, 2005]); to the construction of physical water spaces (the private middle-class home, bath houses, and spas [ibid.]); and to infrastructure which served some (modern citizens) and neglected others (in need of development) (Kooy and Bakker, 2008). Within the frame of the hydraulic mission such infrastructures, together with the design of urban environments as a whole, can be viewed as means of governing.

The second paradigm, which is a more coherent system of thought, is termed *neoliberal hydromentality* in this book. The term neoliberalism should be understood not as a political ideology but as a practical art of government. Such an understanding makes possible a problematization not only of market solutions to water resource governance, but also of its different techniques of governing. Those techniques include, for example, responsibilization and the 'conduct of conduct', that is, the steering of people's behaviour around water. Furthermore, comprehending neoliberalism in this way places a focus on how the agency of water users is utilized to secure the objectives of government (cf. Dean, 1999, p. 15).

In addition, the tensions between market rationalities as solutions to water governance and the interventions required to guarantee the needs of humans and ecosystems, which are central features of global water governance, can be viewed as part of the (neo)liberal problematique of governing without governing 'too much' (Foucault, 2008, p. 319), or too little. This means that governing approaches that discursively lend themselves to inclusion and rights, and which view water as a 'public good', are to be understood as neoliberal in a governmentality sense.

Such insights are central to governmentality studies of water governance. In cases such as Ecuador (Boelens *et al.*, 2015), India (Babu, 2009; Birkenholtz, 2009), and South Africa, inclusionary approaches to water governance have been employed to promote the interests of the (local) state. In this literature, we also learn how subjectivities, of both decision-makers and water users, are created in specific hydropolitical relations of power (Birkenholtz, 2009; Hansson, 2014; Hellberg, 2014, 2017), and we learn about the role that identities play in relation to water infrastructure and access to water (Kooy and Bakker, 2008). Another insight is into how practices of resistance are central to the dynamics of different hydromentalities (ibid.; Birkenholtz, 2009; Boelens *et al.*, 2015).

In exploring the mentalities, rationalities, and techniques of the 'hydromentality' of eThekwini, we glimpse how water management, as an art of government, makes distinctions between different populations, with regard to how certain populations are targeted as suited to particular technological solutions, and how they are perceived in relation to water scarcity and the wastage of water. We also learn how water users are increasingly invited to get involved in water issues in the municipality, both in terms of decision-making regarding their own service delivery, and in formalized focus groups, user forums, and water-training projects.

In this particular hydromentality, we also see how understandings of water that are different from the dominant one are 'othered' and hence rendered 'less real or less legitimate' (Linton, 2010, p. 11). These mechanisms are concretized in awareness programmes and education interventions, a dynamic which has also been shown in Ecuador (Boelens *et al.*, 2015) and India (Babu, 2009; Birkenholtz, 2009).

Biopolitical 'effects'

To understand the effects of such an arrangement of water service delivery as described above, this book has explored water users' narratives on water, 'water places' (cf. Gibbs, 2009), and water technologies (although any snapshot produced of such dynamic processes can only be incomplete).

What we learn from these narratives is that water is not only related to life in the sense of survival, but also that it is explicitly connected to how people perceive their own and others' quality and way of life, and to how they view themselves in relation to social hierarchy. In that sense, through the water users' accounts we learn about the meaning of the productivity of water. Hence, water's relationship to life extends far beyond sustaining life (*zoë*) and is fundamental to the construction of the proper or the good life (*bios*) (Agamben, 1998) and the good society. How water users tell stories of water is also indicative of ambivalence in relation to the notion of life and what such a notion means in terms of the right to water.

As we recall from the previous chapter, in the water users' narratives, there is a strong adherence to the notion of a universal right to water for *basic needs*. The biopolitical claim that everyone should have the right to water because of it being necessary for survival, often appears as a straightforward one and is adhered to as a principle of equality. However, payment as a way to allocate water above the FBW level complicates the water users' narratives in this regard. In the stories from different communities and different socio-economic backgrounds, the narrators represent payment as a legitimate way of allocating resources, although sometimes in an apprehensive manner. Thus, the narrators themselves produce the distinction that is made between those who rely on water for basic needs and those who can access more if they can pay for it.

To complicate matters further, many of the narrators understand their individual rights in relation to history and how they are used to living their lives. On the one hand, this can present the narrators with a means of dealing with the past and the injustices of the apartheid system through claiming the right to water for everyone. On the other hand, the way that the colonial history of South Africa has continued into the post-apartheid era, through inequalities and differences in lifestyle, can also have the effect of making social hierarchies appear 'natural' in that some water users (especially in the middle-class areas) connect their identities to certain rights and lifestyles. In such stories, issues relating to who is entitled to water, how much, and through what level of service, are understood differently in relation to different populations. Relatedly, in these narratives distinctions are made in terms of Self and Other. For example, in the suburbs narratives of water subjectivities take shape through people contrasting themselves with less fortunate others, which in turn illustrates how a certain form of exotification of others plays a role in the construction of one's own lifestyle, identity, and relationship to water. In biopolitical terms this can be understood as a way of producing images of the difference between 'qualified life' (Wrangel, 2014, p. 185) and 'mere' life. At the same time, however, some narrators refuse to draw a line between basic water and other uses in their accounts. In such stories, the claim to water is one that *exceeds* essential use and instead involves the demand for a dignified and decent life and for political inclusion.

Since some water users position themselves as victims of inequality and others as fortunate and privileged, the production of subjectivities reflects a hydromentality that divides populations and individual water users in terms of what they can pay for and/or are eligible for. Even though it can be argued that these inequalities were, in many ways, established in the colonial and apartheid eras, what is produced in the meeting between governing structures and the water users is something new. In current narratives of water, notions of, and claims to, citizenship and rights, are conflated with the understanding that water is (still) unequally distributed through a system that differentiates between different water users and communities. As a consequence, in some narratives, there is ambivalence in terms of how the self is understood in relation to inclusion or exclusion from the political community. In this process, intersectional subjectivities (rather than race), and the understanding of oneself as responsible and/or sustainable, feature through the ways that the water users understand their water services and the reasons why they get a particular water service (as well as their prospects for being able to live a good life in the future).

Moreover, when exploring notions of water waste/water conservation in the water users' stories, we learn about how water users, through the theme of water, narrate themselves as ethical subjects. Through their articulations of how they view appropriate behaviour in relation to water, they construct notions of responsibility and sustainability, as well as constructing themselves as (un)sustainable and (ir)responsible.

There is a variation between (but also to a certain extent within) the different communities, with regard to the reasons why water users express the fact that they need to care about how much water they are using. In the two communities where people exclusively rely on FBW, the need to be frugal in water use is related to other people's ability to access water in the local community; while in the suburbs there is a strong discourse around the need to care for the environment. In the townships, narratives around notions of water waste and the need to care for one's own consumption function so as to produce responsible economic subjects.

In turn, these different reasons for caring about one's water use are connected to different notions of what water waste is. There is a strong discourse around waste being related to payment, in the sense that non-payment generates wasteful usage, which is in accordance with the dominant discourses on sustainable development, water governance, and the particular hydromentality in eThekwini. Thus, we can see that some narrators construct themselves as sustainable (and responsible) in relation to their ability to pay. There are, however, exceptions to this way of framing desirable conduct.

This comes to the fore in narratives about the, so-called, illegal connections. In some stories these connections are understood as a socially acceptable way of fulfilling everyday water needs, rather than as a form of theft. In other narratives, they are projected as an explicit strategy of resistance against the dividing logics and practices of the municipality. Such stories critique the governing rationales and their dividing effects on the communities, as well as their steering of people through education and responsibilization – techniques that especially target the poor. Here, there is a problematization of how governing logics portray water use and water users in poor communities as a problem, rather than focusing on the quantities of water used in more affluent areas.

Many of the narrators have another view of water waste to the one discussed above. They relate water waste to the quantity of water being used, regardless of whether it is paid for or not. According to such an understanding of waste, the ways in which hydropolitics works create differences between life(styles) in the suburbs, which *can* waste water, and life(styles) outside, which *cannot*. In this regard, class is absolutely central to understanding how the water users produce themselves as responsible and/or sustainable. In poorer households it is more or less a matter of course not to use large amounts of water, but in more affluent ones, it can be understood more as a question of self-presentation and (a desire for) self-reformation to speak about water conservation.

Relatedly, by taking into account the narrators' actual water use (through looking at their bills) in relation to how they construct themselves as sustainable and as carers for the environment (through speaking about how they restrict, or ought to restrict, their water consumption) highlights a discrepancy between values and practices. By understanding this in relation to governing logics, we can see how this discrepancy is beneficial, as the municipality collects more money from high-volume end users. From this perspective, it is these users who make the system work; they can be regarded as both sustainable and responsible, as their consumption of relatively large quantities of water makes it possible to cross-subsidize poorer households' access to water.

However, this logic does not fully override the water users' stories. Many narrators express a sentiment that a high consumption of water is a problem, as these water users see themselves as part of a larger redistributive system in which water availability is limited and scarcity is looming. Guilt in relation to one's own usage is not uncommon among people living in the suburbs. Thus, some of the narrators seemed to construct themselves as ethical subjects through narratives of guilt; inscribing guilt with a productive role can be understood as a way for the water user to show that he or she knows right from wrong in terms of water use (with or without the intention of changing practice).

The point here is not merely that there is a discrepancy between values and practices, but that such a discrepancy signals a particular way of understanding the (un)ethical dimension of water distribution. The design of the system not only creates an unequal access to water and everyday water struggles for poorer communities and individuals; it also creates unease towards those who can access unlimited amounts of water. In this sense, these water users' understanding of a just water allocation does not correspond to governing logics in water management. This offers political potential for a different kind of hydromentality and for resource distribution in general. In such an alternative view, entitlements to water are separated from the ability to pay.

While previous governing discourses have used scarcity to instil a sense of limitation and, in practice, have functioned as a means to govern those who can least afford to pay for water, the recent drought in South Africa has implied a material urgency. Such an experience has the potential to be a catalyst for change in terms of how that water is used and distributed. Along with the rationalities above, a focus on the physical amounts of water that are being used, rather than on revenue collection, can be a way forward, not only for finding ways of addressing ecological sustainability but also for reaching a more socially sustainable mode of water governance.

Studying 'bionarratives'

This book not only invites us to rethink water governance theoretically. It also requires a rethink of how we study water governance (see also Hansson *et al.*, 2015; Hellberg, 2015). In line with Rutherford's critique of governmentality studies (2007) this book has refrained from approaching the hydromentality of eThekwini as a completed project, and instead has treated the success and effects of this particular way of governing as an empirical question to be examined. It has done so by combining a governmentality/biopolitics approach with narrative inquiry. Through its emphasis on temporality, a narrative methodology allows us to complicate the water users' views on service delivery, since associations with the past, and expectations of (or desires for) the future, are always inscribed in their accounts of the present. While people's 'stories' might be seen as far removed from technologies of rule, they are a useful avenue for getting at the biopolitical effects of water governance, that is, the (notions of) lifestyles and subjectivities that are created as outcomes of government. Furthermore, this study has, through its development of the sets of questions that have been posed to the narrative transcripts, provided a tangible approach to how to conduct a biopolitical reading of texts.

The methodological approach has emphasized how different populations and individuals are being governed and how they respond differently to such rationales. As we will recall, the conduct of the water users is far from always being congruous with governing rationales and, due to the agency of the water users themselves on how they form rationalities and subjectivities within relations of power, their conduct is always beyond the full grasp of calculated rationalities of government.

This uncertainty has been explored in relation to how the water users individually make sense of, and (to a limited extent) act, in relation to governing logics. The results of this study also indicate that there is a certain lack of alternative visions for water distribution (apart from the dominant neoliberal one). As we saw, many narrators adhere to dominant stories of water management, including the importance of payment and basic water as a right. Some narrators who emphasize their (previous) activist identity seem to lack blueprints for other just ways of distributing water. Also, in such narratives, basic water features as a progressive idea and payment as a proper mechanism of redistribution. Such an effect of governing logics illustrates the force of neoliberal notions of the organization of society, in that they present themselves as a truth (in turn obtained by market mechanisms).

Finally, in relation to methodology, the messiness and complexity of empirical data, as displayed in the empirical section of this book, make it hard to conduct a tidy analysis of power and of the subjectivities of governing strategies. Nonetheless, this should not be seen as an analytical weakness. Rather, such an analysis focuses on the uncertainty of governing and problematizes assumptions about the effects of power and about the subject in the governmentality and biopolitics literature; people – and their stories – do matter when it comes to our understanding of power and the effects of government.

(Beyond) neoliberal biopolitics

With regard to a more general understanding of the workings of neoliberal governmentality, it has previously been argued that one of the general features of neoliberal techniques of governing is 'the will to divide' (Walters, 2004).

This book has queried such dividing practices and their effects in relation to water governance. More specifically, it has examined how lines are drawn between different populations and individuals regarding the water services they are entitled to or can pay for. The most important line drawn in this context is that between basic water and water above that level. Basic water is a forceful idea that functions to legitimatize payment on policy and management levels. It is also a strong idea in the water narratives in this book. As noted above, the drawing of such a line can also be refused.

Whether or not we draw a line between essential uses and other uses of water in relation to different populations, it is not possible to distinguish the biological work that water does in keeping bodies alive from the political work it does that reaffirms people's sense of self and their place in society. This is a basic conclusion we can draw from the insights developed above, in terms of the importance of water not only for survival but also for how people view themselves in relation to others and to governing rationales.

If such a line is drawn (as it is in dominant discourses on sustainable development and water management), it is productive of exclusion and of the idea of certain populations a 'bare life'. Therefore, with specific regard to the discussion of the right to water – which looms large in water governance literature – this book has explored how we can understand the demand for *basic water* (irrespective of the exact amount), along with a commodification of quantities above that level, as producing or further entrenching distinctions between different forms of life. From such a perspective, we should question the drawing of this line to make way for an inclusive access to water, rather than focusing on finding appropriate figures for an amount of basic water, or on how to design welfare-maximizing tariff structures.

What is at stake here is how the larger discourse on sustainable development, through its emphasis on basic needs, produces a notion of a population that is included in the system of rights (through the right to uphold survival) but is still excluded from the privilege of living a good life. By continuing to explore how neoliberal modes of governing produce relationships and distinctions between mere life and the good life, we can create an understanding of the political paradoxes of (neo)liberalism in terms of its ambivalent relationship to life and the exclusions that neoliberal governmentality necessarily involves. By doing so, and by exploring alternatives to these logics, we can better grasp the possibilities of comprehending a beyond of neoliberal modes of governing. This means that we get rid of the legacy of Malthus, or rather, that we reverse its logics, and instead place responsibility and regulation on the affluent populations (mainly in the Global North), who are the ones who really live without a sense of social and ecological futurity and who are thus locked into an endless expansion, assuming abundance and creating destruction, and thereby being the true catastrophic form of life.

References

Agamben, G. (1998) *Homo Sacer: Sovereign Power and Bare Life*. Stanford, CA: Stanford University Press.

Babu, A. (2009) 'Governmentality, Active Citizenship and Marginalisation: The Case of Rural Drinking Water Supply in Kerala, India', *Asian Social Science*, 5(11), 89–98.

Bakker, K. (2010) *Privatizing Water: Governance Failure and the World's Urban Water Crisis*. Ithaca, NY: Cornell University Press.

Bakker, K. (2012) 'Water: Political, Biopolitical, Material', *Social Studies of Science*, 42(4), 616–623.

Birkenholtz, T. (2009) 'Groundwater Governmentality: hegemony and technologies of resistance in Rajasthan's (India) groundwater governance', *The Geographical Journal*, 175(3), 208–220.

Boelens, R., Hoogesteger, J., and Baud, M. (2015) 'Water reform governmentality in Ecuador: Neoliberalism, centralization, and the restraining of polycentric authority and community rule-making', *Geoforum*, 64, 281–291.

Butler, J. (2004) *Undoing Gender*. London, UK: Routledge.

Dean, M. (1999) *Governmentality: Power and Rule in Modern Society*. Los Angeles, CA: Sage.

Dean, M. (2015) 'The Malthus Effect: population and the liberal government of life', *Economy and Society*, 44(1), 18–39.

Duffield, M. (2007) *Development, Security and Unending War: Governing the World of Peoples*. Cambridge: Polity.

Esposito, R. (2008) *Bíos: Biopolitics and Philosophy*. Translated by Timothy Campbell. Minneapolis, MN: University of Minnesota Press.

Foucault, M. (2008) *The Birth of Biopolitics, Lectures at the Collège de France 1978–1979*. New York, NY: Palgrave Macmillan.

Gibbs, L. M. (2009) 'Water places: cultural, social and more-than-human geographies of nature', *Scottish Geographical Journal*, 125(3–4), 361–369.

Hansson, S. (2014) 'Dependence and Autonomy: Conditions of Possibility for Nigerien Ownership in the Water Sector', *Journal of Intervention and Statebuilding*, 9(2), 256–277.

Hansson, S., Hellberg, S., and Stern, M. (eds.) (2015) *Studying the Agency of Being Governed: Methodological reflections*. London, UK: Routledge.

Hellberg, S. (2014) 'Water, life and politics: Exploring the contested case of eThekwini municipality through a governmentality lens', *Geoforum*, 56, 226–326.

Hellberg, S. (2015) 'Studying the governing of lives through bio-narratives', in Hansson, S., Hellberg, S., and Stern, M. (eds.) *Studying the Agency of Being Governed: Methodological reflections*. London, UK: Routledge.

Hellberg, S. (2017) 'Water for survival, water for pleasure – A biopolitical perspective on the social sustainability of the basic water agenda', *Water Alternatives*, 10(1), 65–80.

Hellberg, S. and Knutsson, B. (2016) 'Sustaining the life-chance divide? Education for sustainable development and the global biopolitical regime', *Critical Studies in Education*. DOI: http://dx.doi.org/10.1080/17508487.2016.1176064

Kaika, M. (2005) *City of Flows: Modernity, Nature and the City*. London, UK: Routledge.

Kooy, M. and Bakker, K. (2008) 'Technologies of Government: Constituting Subjectivities, Spaces, and Infrastructures in Colonial and Contemporary Jakarta', *International Journal of Urban and Regional Research*, 32(2), 375–391.

Linton, J. (2010) *What is Water? The History of a Modern Abstraction*. Vancouver, BC: UBS Press.

Luke, T. W. (1999) 'Environmentality as Green Governmentality', in Darier, E. (ed.) *Discourses of the Environment*. Malden, MA: Blackwell Publishers.

Rutherford, P. (1999) 'The Entry of Life into History', in Darier, E. (ed.) *Discourses of the Environment*. Malden, MA: Blackwell Publishers.

Rutherford, S. (2007) 'Green governmentality: insights and opportunities in the study of nature's rule', *Progress in Human Geography*, 31(3), 291–307.

Tellmann, U. (2013) 'Catastrophic Populations and the Fear of the Future: Malthus and the Genealogy of Liberal Economy', *Theory, Culture and Society*, 30(2), 135–155.

Walters, W. (2004) 'Secure Borders, Safe Haven, Domopolitics', *Citizenship Studies*, 8(3), 237–260.

Wolfe, C. (2012) *Before the law: humans and other animals in a biopolitical frame*. Chicago, IL: University of Chicago Press.

Wrangel, C. (2014) 'Hope in a time of catastrophe? Resilience and future of bare life', *Resilience: International Policies, Practices and Discourses*, 2(3), 183–194.

Index

For Product Safety Concerns and Information please contact our EU representative GPSR@taylorandfrancis.com
Taylor & Francis Verlag GmbH, Kaufingerstraße 24, 80331 München, Germany

www.ingramcontent.com/pod-product-compliance
Ingram Content Group UK Ltd.
Pitfield, Milton Keynes, MK11 3LW, UK
UKHW011013180926
475701UK00004B/719

* 9 7 8 0 3 6 7 5 0 8 3 9 5 *